NON-POSTULATED RELATIVITY

A New Way of Presenting Special Relativity

LEV G. LOMIZE
with Andrei Lomize

Translated from Russian by Associate Professor
Minna M. Perelman

CLUES
Ann Arbor, Michigan

Copyright © 2004 by Lev G. Lomize, Andrei L. Lomize

All rights reserved

Printed in the United States of America

Library of Congress Control Number: 2004094085

ISBN 0-975-55210-4

Clues
Lev Lomize 2461 Sandalwood Circle
Ann Arbor Michigan 48105
(734)327-4769 LevLomize@hotmail.com

Additional address:
Andrei Lomize 2866 Leslie Park Circle
Ann Arbor Michigan 48105
(734)665-2237 almz@umich.edu

I dedicate this book to the memory of
VALENTIN ALEKSANDROVICH FABRICANT
and
KONSTANTIN MICHAILOVICH POLIVANOV,
my unforgettable teachers

Contents

Introduction   11
The message of this book   15

Part 1:
Electromagnetic field as a gateway to Einstein's admirable world

1.1. Electric field   18

How the electric field is defined and why we are so sure of its existence

1.2. Magnetic field and Lorentz force   21

How the magnetic field is defined and how it differs from the electric field

1.3. The law of electromagnetic induction   34

which tells us that the electric field may be created not only by electric charges but also by variations of the magnetic field with time

1.4. Displacement currents and electromagnetic wave   38

which tells us that the magnetic field can be created not only by electric currents but also by variations of the electric field with time

1.5. Maxwell equations and Lorentz transformations   44

where we will come across the Lorentz transformations – a mighty tool for finding the solutions of Maxwell's equations, which govern the behavior of electromagnetic systems

1.6. Electromagnetic field of a charge moving at a constant velocity   49

where the first real step is made toward non-postulated relativity by deriving and explaining the Lorentz contraction of the associated field of a point charge moving at a constant velocity

1.7. What Einstein did to the Wonderland of Lorentz   59

where we try to show how Einstein changed the physical interpretation of the Lorentz transformations and how Einstein's revolution looked like in the eyes of the contemporaries

Part 2:
Space, time and the relativity of motion

2.1. Contracting rods   76

where we explain what makes a solid body contract after being set in motion with a constant velocity through the ether and why a moving observer fails to notice that contraction

2.2. Inertia, energy and their unlimited growth   95

where we will see that an almost weightless Ping Pong ball, endowed with a good deal of electric charge, will acquire a considerable mass as soon as it is set in motion with a constant velocity. That mass will depend on the speed of ball's motion

2.3. Stopping time   111

where we will see that every clock, whatever its design might be, has its own good reasons to slow down its ticking after it is set in motion with a constant velocity

2.4. Time zones on a speeding platform   125

where we will see that an array of spatially separated clocks which are arranged in a single file and whose hands are synchronized with each other by anything you like (e.g. by light signals) have a good reason to lose or gain time according to their positions in the file as soon as they are set in motion with a constant velocity

2.5. Moving against the ray of light   151

where, being fixed to the ether, we will watch how the observer, moving through the ether with a speed v, against the ray of light, propagating with speed c, will measure his or her speed with respect to the ray and arrive at a surprising result: his or her speed with respect to the ray will prove c and not $c+v$ as might be expected from the first sight. It is the instruments moving together with the observer and distorted by motion through the ether that are responsible for that curious result

2.6. Which meter stick is more reliable - that at rest or that in motion?   **160**

where the ether fades away and we at last are left face-to-face with inevitability of Einstein's postulates

2.7. Einstein's postulates   **170**

where Einstein's postulates are given together with some tips about their usage

2.8. Electrification of currents   **194**

where we will see that a neutral current-carrying loop has a good reason to acquire electric polarization after being set in motion with a constant velocity

2.9. The curved emptiness   **209**

where, being in search for an inertial frame of reference, we will visit the space and meet free-fallers – the residents of the space – who will explain to us a lot of interesting things about gravitation, which has been thoroughly avoided throughout the previous part of the book

2.9.1. In search for an inertial frame of reference   209
2.9.2. Meeting with free-fallers   211
2.9.3. Returning to the Earth   217
2.9.4. A mysterious tickle   220
2.9.5. Curved spacetime   224
2.9.6. Summary   231

Conclusion   **232**

Birth and evolution of non-postulated relativity **237**

A historical review

1. How classical physics and special relativity found themselves in opposition to each other   237
2. What Einstein thought about it   239
3. Groping ways to non-postulated relativity   252
4. Building bridge to classical physics   254
5. Closing the gaps   258
Literature   270

If you want to learn more about relativity   **273**
Acknowledgements   **275**
Index   **279**

Introduction

Albert Einstein, in his *"Autobiographical Notes"* published shortly before his death, made a very important remark concerning special relativity, which had been created by him 46 years earlier. He called it "inconsistent" and unusual that the properties of rods and clocks "emerge" from his postulates instead of being normally derived from the equations of mechanics and electricity. Einstein regarded such "inconsistency" as temporary *"with the obligation, however, of eliminating it at a later stage of the theory."* [1] The endeavor of the book you are reading is in accord with that obligation. Space and time are identified here with the properties and behavior of the meter sticks and clocks described by the laws of classical physics. At the beginning of our presentation, these instruments are even thought of as the ones moving through the ether, that later on gradually fades away as the reading and understanding are in progress.

Such a non-traditional approach to special relativity has at least two advantages:

1. Relativistic effects are given as integrated into the pre-relativistic physics so firmly that no change in the way of thinking is needed to understand them. Without the traditional opposition of relativity to the pre-Einstein physics, the comprehension of relativity becomes only deeper.

2. The reader learns not only the relativistic result, but also the dynamics of its origin, which is, strictly speaking, beyond the limits of Einstein's postulates, though never contradicts them. Sometimes this turns out to be helpful even from a purely practical point of view.

---

1. See full citation and other details in the historical review "Birth and evolution of non-postulated relativity" at the end of this book (See page 237.)

But dear reader, let me give you an important warning: If your only intention is to memorize the relativistic effects together with their short and straightforward derivation from Einstein's postulates, then this book is not for you. You can use a lot of very good traditional books about relativity. But if, on the contrary, you are striving for the physical mechanisms, responsible for the relativistic "miracles", you have no other option but to read this book and other non-numerous associated literature – pioneered by L.Janossy and E.L.Feinberg – given in a historical review "Birth and evolution of non-postulated relativity" appended to this book (See p.p. 237-272.)

Despite the seeming simplicity of the text and elementary mathematics, this book may appear not so easy. Time and again, you will have to get to the point of understanding some phenomenon connected with motion. Though you will be doing it under the guidance of an author, it may take your time and effort to touch the roots unless you are an expert in physics. But the harder you study, the simpler the ultimate result, which will be memorized just by itself and could be explained to anyone in very simple terms (often even without any mathematics). After such excursions, your confidence in special relativity will become firmer, and all of your doubts in its validity and reality, if any, will fade away.

Now a few words about the author. Born in 1931 in Tbilisi (Georgia, the former USSR), Lev Lomize resided in Moscow from 1949 to 1997. Then he moved to Ann Arbor, Michigan, USA. In 1955, he graduated the Moscow Power Institute, where he studied physics and electrical engineering under the guidance of professors K.M.Polivanov and V.A.Fabrikant. After graduation, Mr. Lomize worked first at the Institute of Radio and Electronics of the USSR and then at the Moscow Institute of Radio Engineers (headed by A.L.Minz) until his removal to the USA. He was engaged then in physical research and engineering connected with developing linear charged particle accelerators (linacs). After receiving the Russian equivalent of Ph.D. in 1961, he took part in developing and launching the 100 Mev injector of the Serpukhov synchrotron in the 1960s, and the Russian analog of Los-Alamos Meson Facility in the 1970s. His publications were devoted to beam loading in linacs and electromagnetic radiation emitted by intensive bunched beams of electrons, such as the transition radiation and the Cherenkov effect. As for his hobbies, let them be reported as a personal story.

R̲elativity entered my life as early as in my childhood when I learned about its existence from the popular book on physics published by Yakov Perelman – the Russian writer who was remarkably effective in making scientific knowledge accessible even to kids. The flattened vehicles from that book, with the flattened passengers, as seen by a resting observer, imprinted into my memory as a riddle to be solved in the future. However, this future stretched for a considerable part of my life. At first, I tried to perceive the special relativity in the traditional way. Though everything was smooth and irreproachable from a formal standpoint, I could not find the answer to the childish questions: "What makes the moving rod shorter and the moving clock slow? If the postulates somehow do it, what is the mechanism they use for doing so?" The first hint came out when I was conducting my early theoretical research on the electromagnetic radiation of a bunched beam of charged particles. [1] With the Maxwell equations taken as a starting point for the derivations, I decided to "simplify" the problem by neglecting relativistic "corrections" which were supposed to be made later on. To my extreme surprise, the relativistic "corrections" turned up from the derivations automatically – just by themselves – as though Maxwell had known about relativity. Neither did Newton know about relativity when formulating his universal laws of motion, which are successfully used all over the world in the computations on beam dynamics of relativistic bunches of particles whose speed is very close to that of light. It was Newton who had formally treated the mass in his second law as though it could depend on the velocity.

Time and again I had a great pleasure to track down the classical way of explaining relativistic effects without using Einstein's postulates. Some of such explanations popped out of my current theoretical derivations almost automatically. As for the others, sometimes it took years of my leisure time to dig them out of the bowels of classical physics. But, as a rule, the longer it took me to comprehend an explanation, the simpler the result was. Later on, I realized that perhaps I was following the way humanity would have been doomed to go if Einstein had been born a cen-

---

1. I did it together with A.N.Vistavkin.

tury later. Anyway, by the early 1980s, I had enough material to share it with a wide audience.

At the first stage of this research, I regarded it as a part of my self-education. But in 1974, I suddenly discovered that I was not alone in my attempts to build a bridge between classical physics and relativity. E.L.Feinberg, one of the best scientific minds in Russia, involved in his research in the field of quantum electrodynamics, spared some of his valuable time to clarify the issue. Supported by V.L.Ginzburg and other theorists from the Lebedev Institute of Physics (Moscow, USSR), in 1974, he published a clear and instructive article devoted to this topic. He kindly lent me the book published by L.Janossy in Budapest (Hungery) in 1971, where most (but not all) relativistic effects were derived from classical physics. This inspired me to turn my casual hunting for the classical roots of relativity into a systematic study. A thorough inspection of the literature revealed a very interesting history of the issue, starting even from Einstein himself. This inspection convinced me of the need for a new special book that would explain the main relativistic effects in terms of classical physics. "Non-Postulated Relativity" has been written as a fair approximation to it. Only the aberration of light and the longitudinal Doppler effect were not included here in order not to distract the reader from the main ideas.

The message of this book

When set in motion with a relativistic speed, every body has its own good reasons to contract its length and to slow down its behavior. Most books devoted to special relativity leave these good reasons in shadow when using Einstein's postulates as a starting point for all the explanations.

In contrast to that tradition, this book is focused on the physical phenomena underlying the relativistic effects. To expose those phenomena, relativistic effects usually are considered here in one and the same inertial frame of reference both before the acceleration and after it, with no attention paid to inconveniences and complications which may be caused by such restriction. The same restriction is imposed when different relativistic effects are considered simultaneously to better understand their physical essence. Using such a "privileged" frame of reference brings us ultimately to the method practised by our ancestors in 19th century when the existence of the ether did not arise any doubts.

It turns out that in our striving for explaining relativistic effects, we inadvertently follow our ancestors and temporarily assume that there is a universal, absolute, all-pervading ether. Unlike our ancestors, we are not concerned too much with the ether's internal structure or design. We only use it as a privileged frame of reference to be eventually given up or replaced by the measuring devices. "Surprisingly", this rejection of the ether turns out to be absolutely painless – leaving all our results unaffected and bringing us eventually to Einstein's postulates. This will be not an easy journey, but we will win a good reward: Instead of memorizing special relativity in a blindfolded manner, we will figure out the real mechanisms responsible for the length contraction and time dilation, including even relativity of simultaneity or time-space dependence.

The journey, as I hope, will not be too boring for us – we will feel like time travelers following the way which science would take if Einstein was born a century later.

PART 1

ELECTROMAGNETIC FIELD AS A GATEWAY TO EINSTEIN'S ADMIRABLE WORLD

1.1. Electric field[1]

How the electric field is defined and why we are so sure of its existence

Two electric charges – q_1 and q_2 – placed a certain distance r apart, interact with each other. Charges of the same sign repel and those of opposite signs attract each other. The force of their interaction F_e satisfies Coulomb's law:

$$F_e = \frac{q_1 q_2}{r^2} . \qquad (1.1)$$

This force is proportional to the product of the charges and to the inverse squared distance between them. The constant of proportionality depends on the units of measurement that are applied to the amounts of the charge, distance and force of interaction. In the CGS system (where centimeter, gram, and second are used as basic units) this factor is equal to unity and therefore is not seen in the formula (1.1). When the charges repel each other, the force is assumed positive, and in the case of attraction – negative.

Coulomb's law may be used as a basis for the following physical definition of a charge. **The charge of a particle is a physical quantity, defined as the force of interaction between two identical stationary particles separated by a unit distance.** In this definition, the concept of force is borrowed from mechanics.

If one of the charges (say, q_1) was suddenly accelerated and flew far away, then, in accordance with the experimental evidence, the charge q_2 would "know" about it not at once but some time later, and until then, the charge q_2 would continue to feel the force F_e

---

[1] Experts in electrodynamics may skip Sections 1.1–1.4, which only serve to prepare the reader for understanding the Lorentz contraction of the field of a charge moving at a constant velocity.

1.1. ELECTRIC FIELD

that would retain its former value determined by the expression (1.1), though, in fact, the charge q_1 would no longer be there. This makes us introduce the concept of an electric field that occupies some region of space and exists independently of the charge which has created that field. A certain physical variable is assigned to every point of such region. This variable is called *the strength* **E** *of the electric field*. This strength, or simply an electric field, at some point of space is defined as a force exerted on a stationary unit test charge placed at that point. The force F_e, exerted on the charge q_2, is equal to the product $E q_2$ and exists as long as there is field E at the point occupied by the test charge q_2. Making use of the concept of an electric field E, we may split formula (1.1) into two independent equations:

$$F_e = q_2 E; \qquad E = \frac{q_1}{r^2}, \qquad (1.2)$$

of which the second one shows how the electric field decreases with the growth of the distance between the charge q_1 and the reference point.

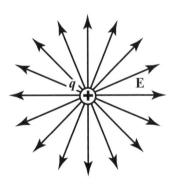

Fig.1. The lines of force of the electric field produced by a single point electric charge resting at a certain point of space.

Because the electric field is defined as a force, it is a vector quantity which tells us not only the magnitude of the field but also its direction. The electric field is often envisioned by means of lines of force. These lines indicate the direction of vector **E**, which is either tangent to the line (if it is curved), or coincident with it (if it is straight). A particular example of such field is given in Fig.1. It is the electric field of a single stationary positive charge. Its lines of force are straight rays originating on the charge and going off to infinity. The magnitude E of this electric field is proportional to the density of lines, which can easily be seen in Fig.1 – as you go farther away from the charge, the distance between the lines increases. This is in agreement with the second formula in (1.2) that prescribes the inverse proportionality between the magnitude of the field and the distance separating the reference point from the charge. All lines of force in Fig.1 have their beginning, but none of them has an end. If a negative charge q_2 is placed at some distance from the positive charge q_1, then the lines of force will get curved and converge to charge q_2. Eventually they reach charge q_2 and terminate there.

The electric field **E** at some point of space is responsible for the force $\mathbf{F}_e = q_2 \mathbf{E}$ which is exerted on charge q_2 placed at that point. Under the action of force $\mathbf{F}_e$, charge q_2 is accelerating and gaining some kinetic energy. At that very moment, charge q_1, that has produced this field, might have left its place for a while and be unable to take any part in that process. Because the energy cannot arise from nothing, we have to admit that an electric field owns some energy. This is one of the forms of potential energy as qualified from the standpoint of mechanics. The density of that energy per unit of volume is proportional to the squared electric field E_2.

1.2. Magnetic field and Lorentz force

How the magnetic field is defined and how it differs from the electric field

1.2.1. Ampere's law and magnetic lines

When charges move, they form an electric current. Current I is defined as an amount of charge passing through the cross-section of a conductor per unit time. A part of the current passing through a unit area of the cross-section is called a current density **j**.

The currents that exist in different regions of space interact with each other. The currents of the same direction attract and of opposite directions – repel each other. The force F_m of interaction between two parallel straight currents I_1 and I_2 obeys Ampere's law:

$$F_m = k\frac{2I_1 I_2}{r}l, \qquad (1.3)$$

According to this law, the force, applied to a unit of conductor's length, F_m/l is proportional to both currents I_1 and I_2, and is inversely proportional to the distance r between the currents. Factor k in (1.3) is a constant that can be found experimentally and depends on the units that have been chosen to measure the values of the current and force of interaction. In the CGS system, it is equal to $1/c^2$, where c is the speed of light in vacuum. Such a coincidence looks very strange because light seems to have nothing in common with the interaction of two stationary currents. If the system of units was changed, the factor k would take another value, but the ratio of the two constants - one in Coulomb's law (1.1), and the other in Ampere's law (1.3), would always remain equal to c^2, whatever units were chosen there. This testifies to a deep internal connection between the Coulomb and the Ampere laws, and makes us suspect that these laws still have something in common with the propagation of light in vacuum. This connection will become less mysterious after our knowledge is broadened in the subsequent sections.

If current I_1 gets suddenly switched off or quickly removed, then, as experiments show, current I_2 would "learn" about it not instantly but some time later, and until then I_2 would not feel any change in the force F_m and would behave as if I_1 were never switched off or driven away. This is a good reason for introducing a concept of a magnetic field with its induction **B**, that is produced by current I_1 in every point of the surrounding space. As soon as this field is created, it can exist and act even in the absence of I_1. Because the induction **B** reflects not only the magnitude of the force F_m, but also its direction, it is a vector.

Man learned long ago how to detect the direction of a magnetic field. This can be done with a compass. When we are going north by a compass, we are moving along a line of the Earth's magnetic field. The vector **B** is tangent to this line everywhere. It is interesting how magnetic lines look like when they are produced by current I_1. Let us replace current I_2 with a compass. What direction will it show? It may seem that the compass must point at the current I_1, i.e along the force which acted upon the current I_2 before its removal. The actual behavior of the pointer proves more sophisticated. It is always oriented along a circle whose center lies on the current I_1 and whose plane is perpendicular to that current. If we go in the direction shown by the pointer, we will have to encircle current I_1 and return to the starting point of our path.

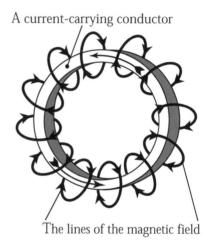

A current-carrying conductor

The lines of the magnetic field

Fig.2. The lines of the magnetic field created by a closed current-carrying conductor.

Thus, the lines of the magnetic field produced by current I_1, have the form of rings threaded onto I_1. The direction of these lines can be found by means of the well-known screwdriver rule – the rotation of the screwdriver which moves along the current indicates the direction of the magnetic lines. The real currents are always curved, because they form a closed circuit.

1.2. MAGNETIC FIELD AND LORENTZ FORCE

Such closed circuit, with magnetic lines in the surrounding space, has a form of a ringed circle shown in Fig.2 where the circle is represented by a closed current-carrying conductor and the role of magnetic lines is played by the rings. The rings are distributed uniformly along the conductor and can make many layers, of which only one is shown in the figure. In the vicinity of the conductor, the magnetic lines are distributed more densely than far from it. This is in agreement with formula (1.3), which prescribes increase of force and, hence, a higher concentration of magnetic lines in the vicinity of the current.

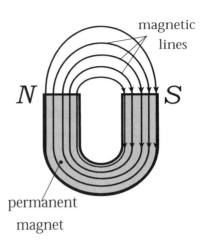

Fig.3. Even in a permanent magnet, the lines of the net magnetic field are always closed. None of them has any origination or end points.

We will see later that magnetic lines may have an elongated shape, but they are always closed. This is their main distinction from the electric lines of force, that may originate and terminate in the electric charges. Even if a magnetic field is produced by a permanent magnet, its magnetic lines, emerged from the north pole of the magnet, do not terminate in its south pole. Instead, they penetrate into the material of the magnet and return to the north pole from within so as to form closed loops as shown in Fig.3.

1.2.2. The Lorentz force

So far, we have been studying how the magnetic field is created. It's time now to see how that field acts upon an electric charge. It should be noted from outset that its action on an electric charge is very different from that of an electric field. If the charge is at rest, it is never affected by a magnetic field, however strong it might be. Even in motion the charge does not feel the magnetic field when its motion is directed along the magnetic lines, and only the charge which moves across the magnetic lines experiences the action of the field. If the velocity **v** of the charge is directed at a certain angle to the magnetic induction **B**, then **v** must be resolved into two components – parallel and perpendicular to the induction **B**. (See $\mathbf{v}_\parallel$ and $\mathbf{v}_\perp$ in Fig.4.)

The force exerted on the charge is determined only by the perpendicular component $\mathbf{v}_\perp$, and does not depend in any way on the parallel component $\mathbf{v}_\parallel$. If at a certain point of space there are both a magnetic field with induction **B** and a charge q moving with a velocity **v**, then the force F_m, exerted on the charge, is proportional to the product of the three variables:

$$F_m = kqv_\perp B, \qquad (1.4)$$

where the constant of the proportionality k is to be found experimentally and depends on the choice of the units. As measurements show, in the CGS system this constant is equal to $1/c$, which is a good reason for introducing a new variable $\beta_\perp = v_\perp/c$ in (1.4)

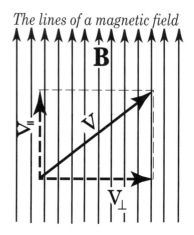

The lines of a magnetic field

Fig.4. Vector **v** is the velocity of a charge crossing the magnetic lines. It is resolved into the components $\mathbf{v}_\parallel$ and $\mathbf{v}_\perp$, of which only $\mathbf{v}_\perp$ is responsible for the force exerted on the moving charge.

$$F_m = q\beta_\perp B \qquad (1.5)$$

Such change of variables means that since now the speed of light is used as a unit of measurement for the charge's speed of motion. Having compared the expressions (1.5) and (1.2) for the forces F_e and F_m, one can see that, in the CGS system, a magnetic field B and an electric field E are expressed in the same units. This remark may prove useful for us in the future.

Formula (1.5) gives the magnitude of the force F_m, but tells us nothing about its direction. Where is this force directed? The answer can be memorized in the following short form – the force F_m, exerted on the charge q, is perpendicular to all vectors that are

1.2. MAGNETIC FIELD AND LORENTZ FORCE

responsible for this force. According to formula (1.5), there are two such vectors – the magnetic induction **B** and the velocity of the charge's motion across the magnetic lines. The force F_m is perpendicular to the both vectors **B** and $\mathbf{v}_\perp$, which are, in their turn, perpendicular to each other. If shown in Fig.4, the force F_m would be directed out of the page.

By now, we are ready to complete the full expression for the net force **F** (both electric and magnetic), exerted on the charge q that moves with a velocity β through a certain point of space, where there is an electric field **E** and a magnetic field **B**:

$$\mathbf{F} = q\mathbf{E} + q[\boldsymbol{\beta}_\perp \mathbf{B}] = q(\mathbf{E} + [\boldsymbol{\beta}_\perp \mathbf{B}]). \quad (1.6)$$

In this formula, the square brackets mean that the two vectors inside the brackets, when multiplied by each other, give a third vector that is perpendicular to both of them. The three vectors – $\boldsymbol{\beta}_\perp$, **B**, and $[\boldsymbol{\beta}_\perp \mathbf{B}]$ – are oriented with respect to each other like the three axes – Ox, Oy and Oz – of an ordinary rectangular coordinate system.

Expression (1.6) (or, more exactly, its magnetic part (1.5)) was proposed more than a century ago by H.A.Lorentz and is widely known as the Lorentz force. This formula, in spite of its simplicity (or rather due to its simplicity), deserves a high esteem. It took decades for physicists to arrive at that formula after the discovery of Ampere's law (represented above by relation (1.3)). In accordance with (1.6) the force, exerted on the charge in a magnetic field, may lie only in a plane perpendicular to vector **B** and, hence, to the magnetic lines. This means that the magnetic lines, unlike the electric lines, are directed not along the forces exerted on the charge, but always perpendicular to them. Nevertheless, they sometimes are also called "lines of force" because their direction determines the planes in which the magnetic forces are acting. Sometimes this conclusion seems to disagree with experiment. If some iron objects are brought near a magnet, they do not try to run somewhere aside. Instead they jump straight to the poles of the magnet, don't they? This gives an impression that the force, exerted on the iron objects, points along the magnetic lines rather than perpendicular to them, which would be in contradiction with the formula for the Lorentz force.

The explanation of this "paradox" is given in Fig.5. A circular current-carrying loop, whose plane is perpendicular to the plane of drawing, is shown there in the neighborhood of a magnetic pole *N*. The plane of this loop is perpendicular to the flux of magnetic lines. This is just the way how interatomic currents are oriented inside of a piece of iron placed into the magnetic field. This figure tells us that the Lorentz forces **F** are not horizontal, but are inclined upward – to the magnetic pole. It is this inclination that is responsible for the force that attracts iron objects to the magnet.

It turns out that the attraction is directed not along the magnetic lines but toward the regions where these lines

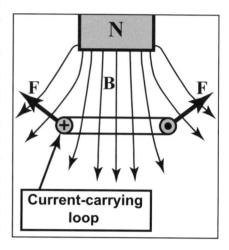

Fig.5. Forces **F** exerted on the current-carrying loop in the neighborhood of pole *N* of a permanent magnet Being inclined to the plane of the loop, these forces not only stretch the loop, but also push the loop toward the pole. The magnetic field of the loop itself is not taken into account.

become denser. If the field were uniform, then the magnetic lines would be parallel to each other and there would be no attraction at all. To make it even more evident, you can do (or just imagine) the following simple experiment. Take an iron rod and insert its end into a coil with a current. The rod will be immediately drawn into the coil because the density of the magnetic lines inside the coil is greater than outside. Now insert the end of the rod into the coil from the opposite side. The rod will be drawn into the coil just in the same way. In the first of these two situations, the magnetic lines are directed inside the coil, while in the second one they are directed outwards. But in spite of it, the rod is drawn inside the coil in either case. The pieces of iron, being attracted to the pole of a magnet, respond not to the magnetic lines themselves but rather to the convergence of these lines. Looking apparently strange, this happens in full accordance with the formula (1.6) for the Lorentz force.

1.2. MAGNETIC FIELD AND LORENTZ FORCE

The Lorentz force (1.6) is of great importance in electrical engineering. The operation of almost all the electric motors and rotating generators is based on the action of this force. The basic idea of a direct current motor is illustrated by Fig.6. The main part of a motor is a rectangular conducting frame that can rotate around the axis OO'. A discontinuity is arranged in one of the sides of the frame to connect the frame with the external source S of the direct current i. The connection is accomplished through a collector which is designed as a short conducting cylinder cut into two halves along its axis. The halves are isolated from each other by a sheet of insulating material. The collector and the frame are fastened to each other comprising a rigid construction (a rotor), rotating around the axis OO'. Two stationary thin elastic strips (the so-called brushes) are sliding over the surface of the collector so as to provide two sliding contacts between the rotating frame and the external source S. The rotor is placed into a stationary magnetic field **B**, which, in the figure, is directed vertically. The field is produced by an electric magnet called a stator (not shown in the figure) and is usually supplied from the same source S as the rotor.

The electric motor acts in the following way. The source S drives an electric current i through the conductors of the frame. The electrons, participating in this current, move along the upper and the lower conductors of the frame in opposite directions, perpendicular to the magnetic field **B**. In accordance with the expression (1.6), there appear two forces $\mathbf{F}_1$ and $\mathbf{F}_2$ which act perpendicularly to both the electron flow (i.e. current i) and the magnetic induction **B**. The current i has opposite direction in the upper and the lower conductors of the frame, and so do the forces $\mathbf{F}_1$ and $\mathbf{F}_2$ which produce a torque exerted on the frame. But this torque reverses its direction every time when the upper and the lower conductors interchange. This difficulty is overcome by means of the collector which rotates between the two stationary brushes and reverses the current so as to guarantee that the Lorentz forces drive the rotor always in the same direction. Under the action of these forces, the rotor would move faster and faster. What constrains this acceleration? If the speed of rotation were not confined, the rotor would fly apart, being destroyed by the centrifugal forces.

Surprisingly, it is again the Lorentz force (1.5) that gives the answer to this question. Indeed, the electrons within the conductors of the rotating frame simultaneously take part in two different motions. On the one hand, they move along the conductors as participants in the current that makes the rotor move. On the other hand, they are having a ride on the frame rotating around axis OO'. (See Fig.6.) Since both these motions are perpendicular to the magnetic induction **B**, they give rise to additional Lorentz forces, that are applied to the electrons and act against the current i. There appears an electromotive force (EMF) in the frame which acts against the external source S. The faster the frame rotation, the stronger this counter-EMF becomes until it finds itself almost in balance with the voltage of the source S. The word "almost" reflects a certain small part of the voltage which is consumed to overcome the electric resistance of the circuit, while the main part of this voltage counterbalances the EMF created by the Lorentz force.

Thus, the rotation of an electric motor and the generation of a counter-EMF are explained by the same reason – by the action of the Lorentz force. But historically, the latter of these two phenomena turned out to be named electromagnetic induction in order to distinguish it from the former one. At that time, it was not yet obvious that both these phenomena were of the same origin. Therefore it is better to refrain from using the term "electromagnetic induction", sparing it for another phenomenon that will be considered in the next section.

Let us now remove the source S from the external circuit, leaving the circuit in a disconnected state. As for the motor, let it be driven now by a turbine, which is not shown in Fig.6. The constant magnetic field should be retained by connecting the stator to some auxiliary low-power source of electric current. The crowd of free electrons within the conductors are having a ride on the rotating frame in the direction perpendicular to the magnetic field. This ride brings about a Lorentz force that drives electrons along the conductors from one of the brushes to the other. Being unable to move anywhere farther, the electrons pile up on one of the brushes leaving the other brush with the same amount of positive charge. There arises a voltage between the brushes, which is continuously growing. Our electric motor has turned into an electric generator. How long will its voltage be growing? What will counterbalance the driving action of the Lorentz force?

1.2. MAGNETIC FIELD AND LORENTZ FORCE

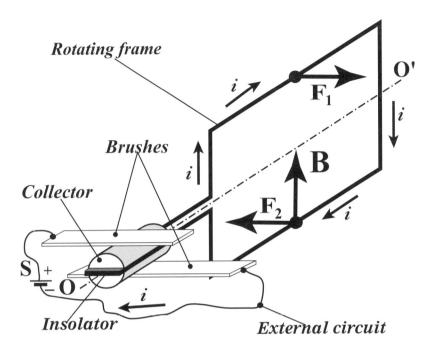

Fig.6. A current-carrying rectangular frame with a collector is rotating around the axis OO' in a stationary magnetic field **B** under the action of the Lorentz forces $\mathbf{F}_1$ and $\mathbf{F}_2$. This outline demonstrates the role of the Lorentz force in electric motors and generators. The stator that creates the magnetic field **B** is not shown in the figure. In the case of a generator, the external supply S is to be replaced by a resistive load.

The answer to this question is given by the formula (1.6) with taking into account the electric field presented there. The fact is that the charges accumulated at the brushes and in the external conductors produce an electric field **E** that acts not only in the external wires (which provides a voltage there), but also within the conductors of the rotating frame. Inside the conductors, this electric field acts against the second (magnetic) component of the net force (1.6). The piling up of extra charges at the brushes ends as soon as the two counteracting items in formula (1.6) cancel each other. In other words, once the equality

$$\mathbf{E} = -[\,\beta_\perp \mathbf{B}\,]$$

is reached, no further redistribution of the electrons will be possible. It follows, for example, from this relation that the output voltage of the generator should be proportional to the speed of rotation of the rotor. The faster rotation (that is, the greater $\beta_\perp$), the stronger the electric field required to balance the magnetic force, and the higher the output voltage between the brushes of the generator.

Now, let us close the external circuit of the generator by connecting the brushes to a resistor that stands for the consumers of electric power. There appears an electric current both in the external circuit and in the rotor. The voltage between the brushes remains unchanged, or almost unchanged, because neither the speed of the rotor nor the magnetic field of the stator suffered any noticeable change – the turbine goes on driving the rotor with the same speed, and the stator continues to provide the same level of the magnetic field. Now our consumers will have their electric bulbs lighted up, their irons, stoves, TV, etc. activated. The more consumers are connected to the external circuit, the lower its resistance, and the higher the current, produced by the generator. How long can this current grow? There must be something that limits the power available from a generator. It is once again that the Lorentz force is to be used in order to provide the answer to this question. This time, when there is a current in the frame, the electrons within the conductors of the frame take part in two different motions simultaneously. They not only are having ride on the rotating frame and thus are bringing about the force by means of which the electric power is generated, but also are moving along the conductors as participants of the electric current. Since this second motion is also directed perpendicularly to the magnetic field, it causes an additional Lorentz force, that tries to stop the rotation or even to override it. If it were not for the turbine, the generator would stop immediately and would cease producing the electric power. The higher the current in the circuit, the stronger this resisting Lorentz force becomes, and the more sizable investment is required of the turbine to overcome that force. If the current goes on growing, then, sooner or later the turbine will get exhausted, its rotation will slow down, the output voltage will drop, and the electric bulbs of the consumers will die out. As we have seen, it is again the Lorentz force that is responsible for all those important phenomena. When the pointer in a compass changes its orientation, it is again the Lorentz force that is responsible for this effect. There is a lot of tiny current-carrying loops inside the pointer, whose role is played by interatomic electric currents. If the pointer is magnetized, most of these frames have their planes oriented perpendicularly to the pointer's axis. When immersed in an external magnetic field, the pointer works just like a rotor of an electric motor. But having no collector, it turns only once and stops in an equilibrium position facing north. If an electric motor were deprived of its collector, it would behave exactly in the same way.

1.2.3. In search for a reference

Formula (1.6) for the Lorentz force is one of the pillars which support classical physics. Therefore, the physical meaning of all its constituents deserves of a thorough examination. Field **E** is defined as a force exerted on a unit charge fixed to *a given point of space*. We were speaking about it on page 19. Field **B** can be defined in a similar way. The magnetic induction (or simply the magnetic field) **B** at *a given point of space M* is a vector, whose direction is shown by a compass placed at point M and whose magnitude is determined by the force, exerted on a unit charge passing through M *at a unit speed* in any direction perpendicular to the pointer of the compass. Although there are many different directions perpendicular to the pointer, it does not matter which of them has been chosen by the moving charge. In accordance with (1.6), the magnitude of the force does not depend on this choice. It does not depend on the sign of the charge either.

The definition given above for the magnetic field, as well as the formula (1.6) for the Lorentz force, are meaningless from the standpoint of physics unless we know the exact sense of the words "the given point of space" and relative to what the velocity of the moving charge must be measured. At the time of Lorentz, everyone believed in the existence of the ether as a certain all-pervading medium that occupies all the universe. Fields **E** and **B** were thought of as some perturbations of this medium. Although the structure of the ether remained mysterious, the very existence of the ether was of crucial importance, because it gave a universal background relative to which the velocity of motion $\mathbf{v}_\perp$ could be counted. "The given point of space" in the definitions of **E** and **B** would also be fixed right to the ether.

Later on, this point of view was radically revised. In his special theory of relativity, Einstein has shown that "the given point of space" must be fixed not to the ether, which successfully evaded all the attempts on detecting it, but to the measuring instruments, that are inevitable participants of all our discussions, definitions, conclusions, and physical laws. All physical quantities, such as magnetic field **B**, the velocity β of the charge, the force **F**, must be measurable to acquire a real physical meaning. If we say that the charge q is at rest or in motion, we imply its being either at rest or in motion relative to the instruments by which it is measured.

A curious reader might have every good reason to regard this idea as somewhat mind-stretching. If the result of a measurement depends on the choice of the instruments involved — whether they are at rest or in motion — then the properties of the instruments become dependent on their velocity. Strange as it is, this idea turned out to be true. Later on, we will learn that all instruments have their own good reasons to behave so strangely. But we are not prepared as yet to grasp these reasons, and we still need some temporary material frame of reference to interpret the quantities which participate in the equation for the Lorentz force. To get out of this vicious circle, we will follow our ancestors and assume that there is a universal, absolute, all-pervading ether. Unlike our ancestors, we will not be concerned too much with the ether's internal structure or design. We will only use it as a hypothetical privileged frame of reference to be eventually given up and replaced by the measuring devices. Surprisingly, this rejection of the ether will be absolutely painless — leaving all our results unaffected and bringing us eventually to Einstein's postulates. This will be not an easy journey, but we will win a good reward — we will figure out the actual roots of special relativity instead of habituating to great ideas coming from the blue. And, as I hope, the journey will not be too boring for us — we will feel like time travelers following the way which science would take if Einstein was born a century later.

In the wake of this assumption, let us now refine the physical meaning of the variables which participate in formula (1.6). Vector $\beta_\perp$ is the velocity of the charge relative to the ether, and fields **E** and **B** as well as force **F** refer to a certain point of that ether. It is important that no other velocity except the velocity of the moving charge is explicitly included into the formula for the Lorentz force. If, for example, field **E** is produced by a moving charge, or field **B** is created by a moving magnet, then the velocity of the charge or the magnet will *not* appear in (1.6) explicitly. The motion of the sources of the field will directly affect only fields **E** and **B**, which determine force **F** in accordance with (1.6). It's just the way how fields **E** and **B** work in physics — as entities which are independent of their

1.2. MAGNETIC FIELD AND LORENTZ FORCE

sources. As for the fields themselves, they are not able to move, because they are nothing else but the local perturbations of the ether. You may wonder — how can it be that the field between the poles of a magnet does not move together with the magnet? It doesn't indeed. It just disappears at one point of space and appears at the next one. That's why we say that the field is propagating — not moving. Let us imagine, say, 20 men lined in a rank. Let each of them have an identical small bright flag in his hand. If these men are raising their flags and then bringing them down in turn, it will then seem to a distant observer that the flag itself is moving above the rank. This impression is quite deceptive. Actually, every flag remains in the man's hand and is not conveyed along the rank. When the flag disappears in some link of the rank, another flag, absolutely identical to it, is raised in the next link. It's just the way how the magnetic or electric field propagates in space.

And what about the lines of force? If a magnet moves, doesn't it mean that the lines of force accompany the magnet? They certainly do. But the displacement of the magnetic lines and the movement of a magnet are quite different kinds of motion. The displacement of magnetic lines is similar to the motion of a sea wave crest on a non-windy day. The crest is certainly moving. But does water follow the wave? Swim beyond the surf line, and you will make sure that the waves which are passing by you in a horizontal direction swing you only in the vertical direction. It turns out that the water does not move anywhere and is just swinging. This is exactly how Maxwell and Lorentz thought of the field of a moving magnet with the all-pervading ether playing the role of the water. This is just the way how we explain the field now, at the start of the new millennium, when we do without ether.

1.3. The law of electromagnetic induction

which tells us that the electric field may be created not only by electric charges but also by variations of the magnetic field with time

In the previous sections, only electric charges and currents served as sources of an electromagnetic field. But as far back as the beginning of the 19th century, Michael Faraday discovered that the electric field can be generated even without any charges or currents whatsoever. Let us take, for example, a beam of circular lines of magnetic field in empty space. Let this lines form a solid ring shown in Fig.7(a). The magnetic lines are enclosed in a circular hood which signifies the volume occupied by the magnetic field. This hood can be

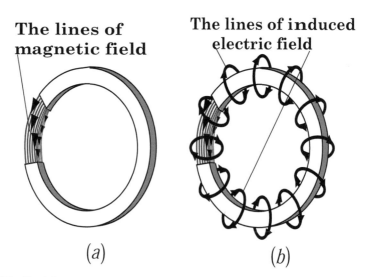

Fig.7. (a) A stationary magnetic field confined within a circular hood which signifies the volume occupied by the magnetic field. No electric field is generated here. (b) The magnetic field is decreasing with a constant rate. A constant electric field is generated, whose lines of force encircle the magnetic flux. Compare with Fig.2 where it was a stationary electric current that was encircled by the lines of a stationary magnetic field.

1.3. THE LAW OF ELECTROMAGNETIC INDUCTION

thought of as imaginary or made out of a non-magnetic and non-conducting material.

Until the magnetic field, as well as the currents which support it (not shown in Fig.7 for the sake of clarity), remain stationary, nothing happens in the surrounding space, as shown in Fig.7(a). Once, however, the magnetic field starts either to grow or to decrease with time, there appear closed lines of an electric field which encircle the ring filled with magnetic field as shown in Fig.7(b). There appears a familiar ringed circle – an exact copy of which we saw in Fig.2. But the cast has changed. The role of the electric current is played by the varying magnetic field, or, more exactly, by the rate of the magnetic field variation. As for the magnetic field whose creation was shown in Fig.2, its role is played now by the electric field. The electric field points either clockwise or counter-clockwise (with respect to the direction of the magnetic field), which depends on whether the magnetic field inside the hood is decreasing or growing. The magnitude of the generated electric field is proportional to the rate of change of the magnetic field. The constant of proportionality is to be measured experimentally. Thus, *a change of a magnetic field creates an electric field just in the same way as a current creates a magnetic field.* This rule is a qualitative formulation of *Faraday's law of electromagnetic induction.*

If an electric field is induced in accordance with this law, its lines of force have neither beginning nor end, and are always closed. Such a field is called *vortical*[1] to distinguish it from *electrostatic* field which is created by charges. Section 1.1 tells us that every line of force of electrostatic field originates from a positive charge and terminates in a negative one. As for the action of an electric field on a charge, it does not depend on its origin and is always determined only by the magnitude of the field and its direction.

To grasp the point of the law of electromagnetic induction, we have to realize that *the induced electric field is caused by a change in the magnetic field – not by the magnetic field itself.* If a positive magnetic field, say, decreases and becomes negative (i.e. changes its direction), then there is no magnetic field at all at the moment of passing through zero, while the rate of changing may be as high as one likes.

---

[1] Sometimes such field is also called solenoidal, circumferential or circuital.

A quantitative formulation of the law of electromagnetic induction is based on the concept of *flux*. If a closed loop is immersed into a magnetic field, then the flux is defined as the number of magnetic lines intercepted by an area confined by the loop. If the magnetic lines are envisioned as a stream of a certain imaginary fluid, then the flux expresses the amount of the fluid passing through the loop per a unit time. A magnetic flux encircled by a loop is defined as the magnetic induction B inside the loop multiplied by the area confined by the loop.

A changing flux of magnetic lines which is encircled by a stationary closed loop – true or imaginary – gives rise to an electromotive force, that is acting in that loop and is determined by the rate of the change in the magnetic flux.

It should be noted that the mentioned rate of change is taken with a negative sign, because the EMF induced in the closed loop always acts against the cause that is responsible for the change in the magnetic flux. This regularity is known in physics and electrical engineering as *Lenz's rule*. For example, a current-carrying loop creates a magnetic field whose lines cross the area encircled by the loop. When the current starts to change for some reason, so does its magnetic field. This gives rise to the EMF which is directed against the current when it is growing, and does its best to support the current when it is falling. If a coil and a bulb are connected in series with a direct current supply to form a closed circuit, then on switching on, the brightness of the bulb will rise gradually. This is explained by the EMF of self-induction, which is generated in the coil when the current and its magnetic field are growing. This EMF acts against the current and must be overridden by the external EMF before the bulb acquires its normal brightness. If, on the contrary, the bulb is being switched off, the current in the coil goes down together with its magnetic field. This causes the EMF of self-induction that does its best to sustain the current and can even cause undesired sparking in the disconnecting switch. According to Lenz's rule, the electromagnetic induction endows electric circuits with the property of inertia. The current and its magnetic field are unable to arise or disappear instantly, because the magnetic field is a carrier of energy that cannot arise or disappear at once. An infinite power would be required for an instantaneous production of any energy, whereas the real

1.3. THE LAW OF ELECTROMAGNETIC INDUCTION

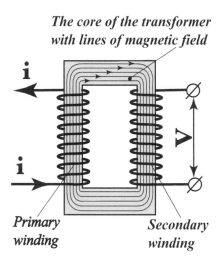

Fig.8. Illustration of how the law of electromagnetic induction works in an electrical transformer. An alternate current i in the primary winding gives rise to the alternate magnetic flux in the core, which in its turn generates an electric field in the secondary winding. This electric field appears as a voltage V at the output of the transformer.

sources of current are able to provide only a finite power, however powerful they might be.

The law of electromagnetic induction is widely used in electrical engineering. The operation of electrical transformers is based on this law. An alternate current in the primary winding of a transformer produces an alternate magnetic flux in the iron core, which in its turn induces an EMF in the secondary winding as shown in Fig.8. It is important to realize that the EMFs in Fig.6 and Fig.8 are caused by different physical phenomena. It is a Lorentz force in a rotating electrical machine (because the conductors are in motion there) and an electromagnetic induction in a transformer (where all the conductors are stationary).

1.4. Displacement currents and electromagnetic wave

which tells us that the magnetic field can be created not only by electric currents but also by variations of the electric field with time

1.4.1. Displacement currents

Thus, in a ray of light which is propagating through emptiness (or through an ether) the electric field is created by variations of a magnetic field. But what creates and maintains the magnetic field itself? Maybe the changing electric field? They are alone in the emptiness, aren't they? If a variation of a magnetic field causes an electric field, why should the inverse process be regarded impossible? Otherwise, the electric and the magnetic fields would not be equal in their rights, would they? The justice was done more than a century ago by James Clerk Maxwell, who proposed a hypothesis that proved a brilliant final link in the classical theory of electromagnetism. He assumed that every change of an electric field at a certain point of space, produces a magnetic field around that point just in the same way as an electric current. In other words, in the ringed circle described in Section 1.2, the role of the electric current can be played (partially or completely) by the lines of force of a changing electric field. Since at the time of Maxwell an electric current was strongly believed to be the only possible source of a magnetic field, the rate of an electric field variation (a partial derivative $\partial \mathbf{E}/\partial t$) was called *"a displacement current"*, being associated with the supposed displacement of some particles of the hypothetical ether. Though later on physicists gave up the idea of those particles as well as even the ether itself – because both of them proved undetectable – the term "displacement current", coined by Maxwell, still remains in use as a reminder of the ether and its hypothetical internal structure.

1.4. DISPLACEMENT CURRENTS AND ELECTROMAGNETIC WAVE   39

When Maxwell applied the term "current" to the rate of change of an electric field, he also multiplied that current by a constant factor that was chosen by him in a very specific way. To understand his choice, it is sufficient to take, for example, a charged non-conducting filament – shown in Fig.9 as a straight gray cylinder – and move it lengthwise in vacuum with a uniform velocity from the left to the right. Because the filament is charged, its motion is equivalent to an ordinary electric current of the same direction. (Since it takes place not within a conductor, it is usually called *a convection current* to distinguish it from *conduction current*.) But this current, if taken alone, is always unclosed. Its snapshot always terminates in the front end of the filament as abruptly as it originates on the rear end. There is an electric field around the filament (not shown in the figure for clarity), that points radially outwards with respect to the filament. Before and after the midpoint of the filament, the lines of electric field (not shown in the figure) are inclined respectively forward and backward. When such filament passes by an observer, the latter can register a change in the electric field. Thus, the electric field, at a certain stationary point of space, is always changing with time, molding a displacement current around the filament, whose

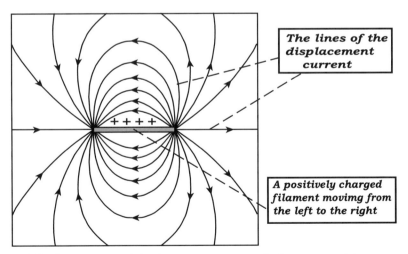

Fig.9. The lines of displacement currents which are created by a positively charged filament moving in vacuum with a constant velocity. This filament, shown as a grey rod, is filled with the lines of a convection current. The lines of the displacement current together with the lines of the convection current make close loops just like the lines of the ordinary current in an electric circuit.

lines are shown in Fig.9. This leads to the situation in which the net current (the sum of the ordinary current and the displacement current) does not terminate at the front end of the filament and splits there into a bunch of lines of the displacement current. Having branched off from the front tip of the filament, these lines turn back toward the rear end of the filament (some of them are directed backward from the very point of their origination), where they are collected at the back tip just in the same way as they originate at the

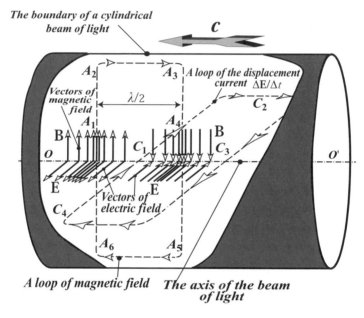

Fig.10. A snapshot taken inside a cylindrical beam of light propagating from right to left with speed c. In the core of the beam, the lines of the magnetic field **B** are parallel to each other. When these lines approach the lateral surface of the beam, they get curved and turn along the beam to form closed loops, one of which is shown in the figure as $A_1 A_2 A_3 A_4 A_5 A_6$. The lines of the electric field **E** are directed horizontally and behave similarly. One of their loops could be obtained by rotating the loop of the magnetic field by 90° round the axis of the beam OO'. For clarity, this loop is not shown in the figure. But it is represented by the relevant loop of the displacement current $C_1 C_2 C_3 C_4$, which makes an interlocking pair with the corresponding loop of the magnetic field $A_1 A_2 A_3 A_4 A_5 A_6$. Due to such an interlocking orientation, the electric and magnetic fields create and support each other while propagating through empty space.

tip. Together with the convection current, they form closed loops just as if they were ordinary electric currents streaming through emptiness. But at the ends of the filament, where the convection current turns into a displacement current (or vice versa), the net current might make a jump if not for the above-mentioned constant factor which was chosen by Maxwell in a special way so as to exclude any jumps.

Though such a choice of the constant factor seemed quite natural, its effectiveness exceeded every expectation. Armed with a displacement current, the 19th century physicists managed not only to display the mechanism of the propagation of an electromagnetic field, but also to calculate theoretically the speed of light c, which proved to coincide with its measured value 300,000 km/s. This was a great success. Look, for example, at Ampere's and Coulomb's laws. Their constants of proportionality were obtained from the experiments that had apparently nothing to do with the propagation of light. Nevertheless, the ratio between these two constants "quite unexpectedly" proved to coincide with the speed of light in vacuum. Such an agreement with the experiment could not be regarded as a matter of chance. On the one hand, it was a brilliant demonstration of the electromagnetic nature of light, and on the other hand, it confirmed Maxwell's daring hypothesis about the displacement currents.

1.4.2. Electromagnetic wave

The discovery of displacement currents paved the way to understanding the mechanism of light propagation. The electric and magnetic fields travel through the world space together. Disappearing in one location, they arise in the next one. This means that at any fixed point of space (or of the ether as was believed at the time) they always change in time. The variation of an electric field at a certain previous point of space generates a magnetic field at the next point, and the magnetic field variation, in its turn, creates the electric field. There arises an electromagnetic wave that consists of interchanging condensations and rarefactions of the field, propagating in space with a speed of light c.

A beam of light is a very instructive example of such electromagnetic wave. Its internal structure is outlined in the snapshot given in Fig.10. Everywhere within the beam – except the region adjacent to its lateral surface – the pattern of the electromagnetic field is very simple. Vectors **B** and **E** are perpendicular to each other. The magnetic lines are all straight and point either upward or downward. The lines of the electric field behave similarly and point either at reader or into the drawing. The wave propagates from the right to the left perpendicularly to both vectors **B** and **E**, i.e. along the axis OO'. The magnitudes of the fields **B** and **E** are represented by the densities of their lines. The distribution of these densities along the axis OO' resembles a sea wave with its crests and troughs. A certain crest A_1 of the wave in Fig.10 corresponds to a condensation of the lines of **B** and **E**. The nearest trough A_4 is also a condensation of field lines which, however, have an opposite direction with respect to the crests. The distance λ between two successive crests – or between any other likewise cross-sections of the beam – is called wavelength and is shown in the figure.

Looking at the rectilinear pattern of the field lines in the bulk of the beam in Fig.10, we may be puzzled – this pattern seems to contradict to what was told in the previous sections about the main property of the field lines – in the absence of free charges, these lines must form closed loops. So they do indeed. To track these loops down, let us begin from the point at which a dashed magnetic line A_1 crosses the axis OO' of the beam – exactly at the middle of the crest of the wave. Being directed upward, this line is at first straight (fragment $A_1 A_2$) and remains straight throughout the cross-section of the beam until it approaches the beam's lateral surface. Being unable to terminate anywhere, it turns right along the surface of the beam (fragment $A_2 A_3$) in search of the way for returning back to the bulk of the beam to form eventually a closed loop. It gets the chance as soon as it comes alongside the middle of the trough (point A_3), where it is escorted down by the downward stream of magnetic lines (fragment $A_3 A_4$). After reaching the opposite side of the beam (point A_5), the line turns left along the beam's surface to come alongside the crest – from which we started tracing it – (point A_6) to be escorted up there by the upward stream of neighboring magnetic lines. This is the last step in creating the closed loop $A_1 A_2 A_3 A_4 A_5 A_6$. According to Maxwell's theory, for this magnetic loop to steadily exist, there must be a flux of a displacement current intercepted by that magnetic line. The central line of that flux – $C_1 C_2 C_3 C_4$ – is shown in Fig.10. It is drawn through the points where the rate of change of the electric field is maximal – exactly midway between the crest and the trough of the wave. It is just there that the water in a sea wave is falling and rising with maximal speed.

1.4. DISPLACEMENT CURRENTS AND ELECTROMAGNETIC WAVE

The rounding of the field lines near the surface of the beam of light makes the field pattern there much more intricate than in the bulk of the beam. Light propagates there not only along the axis of the beam, but partly sideways, causing some divergence of light. The longer the beam, the greater its divergence, however parallel the beam was formed initially. This phenomenon is of a rather complicated nature and is called diffraction. We shall not dwell on it here. Now, that we know how the displacement current works, it's high time to proceed to the next section.

1.5. Maxwell equations and Lorentz transformations

where we will come across the Lorentz transformations – a mighty tool for finding the solutions of Maxwell's equations, which govern the behavior of electromagnetic systems

After Maxwell introduced the concept of a displacement current, the classical electrodynamics became a self-consistent and elegant branch of science. All the newcomers took their proper places in the physical theory – an electric field is produced by either electric charges or variations of a magnetic field; a magnetic field is created by either electric currents or variations of an electric field; an electric current is caused by the motion of electric charges; the fields act on the charges through the forces, determined by formula (1.6) for the Lorentz force. All these ideas were mathematically formulated in the Maxwell equations.

These equations can be used to determine the field vectors **E** and **B** at every point of space x, y, z for every moment of time t, given we know the position and motion of every charge that is responsible for producing that field. These positions can be represented in terms of a charge density $\rho(x, y, z, t)$ per a unit of volume. In a simplified form, the motion of the charges can be introduced in terms of the velocities $\mathbf{u}(x, y, z, t)$ of their motion at every point of space and at every moment of time. Vector **u** tells us about the presence of an electric current whose density **j** is unequivocally determined in terms of **u** and ρ: $\mathbf{j} = \rho\mathbf{u}$. The higher the charges' density ρ and the faster the charges move, the higher the current density. If there are some points of space where $\mathbf{u} = 0$ (i.e. the chargea are at rest), then there is no current there. The case with charges of different velocities concentrated at the same point of space is out of the scope of our consideration. In the Maxwell equations, the partial derivatives of the vectors **E** and **B** are expressed in terms of ρ and **j**. The process of solving these differential equations is not considered here. But for our particular aims we will be able to do even without the Maxwell equations if only we use just one their remarkable property, discovered by Lorentz and expressed in rather simple algebraic equations. Lorentz made his discovery while solving the following problem.

Let us imagine a certain set of charges and currents which are initially at rest. Suppose that their electric and magnetic fields are found as a solution of Maxwell's equations. We want to see how these fields will look like when the entire system of charges and currents is set in motion with a constant velocity **v**. To find the solution of this problem, it would be necessary to solve the Maxwell equations anew. In principle, this is possible. But doing so for the moving sources is much more difficult than for the same system at rest. Even in a simple situation, it may be difficult to solve the equations unless a special procedure proposed by Lorentz is applied.

The essence of that procedure can be illustrated with a simple mathematical example. Suppose the following equation is to be solved:

$$x^8 + 7x^4 - 8 = 0; \qquad (1.7)$$

This is an eighth degree equation. In general, it can be solved only by numerical computations. But there is an extenuating circumstance here that serves well if we guess to introduce a new variable y which is connected with the original variable x in the following way:

$$y = x^4. \qquad (1.8)$$

The substitution of y for x^4 into (1.7) brings us to quite an ordinary quadratic equation:

$$y^2 + 7y - 8 = 0. \qquad (1.9)$$

Solving this equation by means of a well-known quadratic formula, we can easily obtain its roots: $y_1 = 1$ and $y_2 = -8$. After that we get the final solution $x = \pm 1$. The second root $y_2 = -8$ is ignored because x^4 cannot be equal to a negative value.

In this example, the success was achieved by means of lucky change of the variable (1.8), which converted the strange equation (1.7) into the standard quadratic equation (1.9). It might be noted that, in accordance with the initial conditions of the problem, the variable y was not wanted there. It was introduced in the process of solving and might be forgotten altogether, once we made use of it. Of course, most eighth-degree-equations cannot be solved like that. But equation (1.7) is an exception to the rule because it has a remarkable property – it reduces to a quadratic equation once its initial variable x is changed for the new variable $y = x^4$.

As was discovered by Lorentz, Maxwell's equations, which are not even shown in this book, have a remarkable property. When applied to a certain moving system S of charges and currents, they have the same appearance as in the case of some other system S', whose charges and currents *are at rest* and whose variables are represented in terms of variables of system S in a very simple and definite way. It is very important that this new system S' is at rest, because to find out the desired vectors **E** and **B** by solving Maxwell's equations for a system in motion is much more difficult than to do the same for a system which is at rest – just as to find the roots of the eighth degree equation (1.7) is much more difficult than to solve the quadratic equation (1.9). To reduce (1.7) to a quadratic equation, we changed the variable by means of the transformation (1.8). To switch the Maxwell equations from the system S which is in motion to the system S' which is at rest, Lorentz developed a set of simultaneous transformations to be given below. They are technically so simple that can be handled even within the scope of this book.

Technically, the procedure of solving the Maxwell equations for the system S involved the following steps. By changing the variables, which was done by means of the equations known now as the Lorentz transformations, he passed from the true (*non-primed* [1]) variables to some other fictitious (*primed*) variables that made no

---

1 The words "primed" or "non-primed" refer to the variables which are supplied or non-supplied with the superscript ' and will be frequently used below.

physical sense. The Lorentz transformations, which convert non-primed (true) quantities into primed (fictitious) ones, are composed in a special way, so that the velocity of the system turns into zero and the primed system S' of charges and currents proves to be at rest, which greatly simplifies the process of solving the Maxwell equations. This brought him to the solution which comprised only fictitious variables. Using his transformations once again, he returned back from primed variables to the true ones, which brought him at last to the desired solution.

The Lorentz transformations to be given below involve the speed v of the collective motion of system S. This speed enters the transformations not directly, but only in terms of the dimensionless parameters β and γ which by definition, are

$$\beta = \frac{v}{c}; \qquad \gamma = \frac{1}{\sqrt{1-\beta^2}}; \qquad (1.10)$$

and will be used very frequently throughout the book. When v approaches the speed of light c, its dimensionless value β approaches unity, while γ becomes infinitely large. For the speeds that are small as compared with c, the value of β is much less than unity, while γ approaches unity from above: γ is never less than unity. The physical meaning of γ will become clearer in Section 2.2

There may be not only charges but also currents that are involved in the moving system S. This means that every charge is permitted not only to take part in the common motion with the velocity $\mathbf{v}$, but also to move relative to other elements of the system or, in other words, to have its own velocity $\mathbf{u}$ that differs from $\mathbf{v}$. Thus, the Lorentz transformation may contain not only the distribution of the electric charges $\rho(x,y,z,t)$, but also the distribution $\mathbf{u}(x,y,z,t)$ of their velocities. Every local charge at a point with a density ρ has its own velocity $\mathbf{u}$. If some charge does not take part in the current, then the equality $\mathbf{u} = \mathbf{v}$ is observed at the point taken by that charge.

The coordinates x, y, z belong in a stationary rectangular frame of reference which has its x-axis aligned with the velocity $\mathbf{v}$. The vectors $\mathbf{E}$, $\mathbf{B}$, and $\mathbf{u}$ are resolved into their components along the axes x, y, z and are denoted, for example, as $E_x, E_y,$ and E_z.

The Lorentz transformations determine the auxiliary, fictitious, primed variables in terms of the similar true, non-primed variables:

$$x' = [x-vt]\gamma; \tag{1.11}$$

$$y'=y; \quad z'=z; \tag{1.12}$$

$$t' = \left[t - \frac{vx}{c^2}\right]\gamma; \tag{1.13}$$

$$E'_x = E_x; \quad B'_x = B_x; \tag{1.14}$$

$$E'_y = (E_y - \beta B_z)\gamma; \tag{1.15}$$

$$E'_z = (E_z + \beta B_y)\gamma; \tag{1.16}$$

$$B'_y = (B_y + \beta E_z)\gamma; \tag{1.17}$$

$$B'_z = (B_z - \beta E_y)\gamma; \tag{1.18}$$

$$\rho = \rho'\left[1 - \frac{uv}{c^2}\right]\gamma; \tag{1.19}$$

$$u'_x = \frac{u_x - v}{1 - \frac{vu_x}{c^2}}; \quad u'_y = \frac{u_x}{\left[1 - \frac{vu_x}{c^2}\right]\gamma}; \quad u'_z = \frac{u_z}{\left[1 - \frac{vu_z}{c^2}\right]\gamma}; \tag{1.20}$$

For the time being, we take all the primed variables just in a formal way (as it was done by Lorentz), without ascribing any physical meaning to them. If the given system of charges and currents is at rest, then $v = 0$, $\beta = 0$, $\gamma = 1$, and the Lorentz transformations become identities, as was to be expected.

The Lorentz transformations have a remarkable symmetry (not noticed by Lorentz and discovered only by Einstein, who successfully used it as a technical basis of relativity). If all primed variables are regarded as given and all non-primed variables as desired, then after solving the system (1.11)–(1.20) of simultaneous equations in non-primed variables, one is brought to new equations which are identical to the original equations with the primed and non-primed variables exchanged and with the speed v being of an opposite sign. It will be obtained, for example, $x = (x' + vt')\gamma$. You might make sure of it for yourself. No more is to be done here than solving two simultaneous first degree equations in two unknown variables.

1.6. Electromagnetic field of a charge moving at a constant velocity

where the first real step is made toward non-postulated relativity by deriving and explaining the Lorentz contraction of the associated field of a point charge moving at a constant velocity

1.6.1. Derivations

The previous section showed us how the Lorentz transformations look like and what they originally were purported to. In this section, we will see how these transformations work. We will use them here to obtain the field of a single point charge q moving with a constant velocity. When we find this field and compare it with the well-known field of a stationary charge, we will learn how the motion of the charge with a constant velocity affects its electromagnetic field. It will be our first step to special relativity.

So, we want to determine the fields **E** and **B** of a point charge q which is moving at a uniform velocity v along the x-axis and passes the point $x = 0$ at the moment $t = 0$. We will see how the Lorentz transformations will succeed in solving this problem almost automatically – in full accordance with the plot given in the preceding section.

To arrive at the field of a moving charge, it is necessary first to switch over to the primed variables or, in other words, to "move" our charge from our real world, described in terms of non-primed variables, into the imaginary world of Lorentz, described exclusively in primed variables. To perform this "move" we will take the true coordinates of our charge

$$x = vt; \qquad y = 0; \qquad z = 0; \qquad (1.21)$$

and substitute them into the transformations (1.11) and (1.12). This will bring us to the following result:

$$x' = 0; \qquad y' = 0; \qquad z' = 0. \qquad (1.22)$$

Such are the fictitious coordinates of our charge in the world of Lorentz. Since they proved independent of time t, the transformation (1.13) will not be needed. Eqs (1.14)–(1.18) will be used later.

The next transformation (1.19) deals with the space distribution of the charge density and is meaningless in the case of a point charge, which cannot be thought of as having any space distribution. There remains only transformations (1.20) into which the components of the net velocity of our charge

$$u_x = v; \qquad u_y = 0; \qquad u_z = 0; \qquad (1.23)$$

are to be substituted to bring about the following data:

$$u'_x = 0; \qquad u'_y = 0; \qquad u'_z = 0. \qquad (1.24)$$

This indicates that our charge does not create any currents in the world of Lorentz

The results of (1.22) and (1.24) show what becomes of our charge after it is "moved" to the world of Lorentz. It is just a point charge q anchored at the origin of the frame x', y', z'. Now we have to determine the fictitious vectors **E'** and **B'** of that charge in the world of Lorentz. However, we already know that solution from Section 1.1. The charge at rest has no magnetic field:

$$B'_x = 0; \qquad B'_y = 0; \qquad B'_z = 0. \qquad (1.25)$$

As for the electric field **E'**, it is determined by Coulomb's law and has a very simple pattern, shown in Fig.11(a). (See page 56.) At some point of space M', the electric field E' is equal to q/r'^2, where r' is a distance from the reference point M' to the origin of the frame, while the direction of this field is determined by the straight line that connects the origin with M'. To be used for a substitution into the Lorentz transformations (1.14)–(1.18), vector **E'** should be resolved into its components. The ratio of each of them (say, E'_x) to the magnitude of the vector **E'** is the same as the ratio of the relevant coordinate (say, x') to the distance r'. Thus,

1.6. ELECTROMAGNETIC FIELD OF A CHARGE MOVING AT A CONSTANT VELOCITY

$$E'_x = E' \frac{x'}{r'}; \qquad E'_y = E' \frac{y'}{r'}; \qquad E'_z = E' \frac{z'}{r'}, \qquad (1.26)$$

where

$$r' = \sqrt{x'^2 + y'^2 + z'^2}. \qquad (1.27)$$

A substitution of $E' = q/r'^2$ and (1.27) into (1.26) yields

$$E'_x = \frac{qx'}{(x'^2 + y'^2 + z'^2)^{3/2}}; \quad E'_y = \frac{qy'}{(x'^2 + y'^2 + z'^2)^{3/2}};$$

$$E'_z = \frac{qz'}{(x'^2 + y'^2 + z'^2)^{3/2}}; \qquad (1.28)$$

which is equivalent to the relation (1.2) given in Section 1.1 in connection with Coulomb's law.

Now that the fictitious vectors **E'** and **B'** are obtained and represented by relations (1.28) and (1.25), it is time to return the charge from the world of Lorentz to our real world. This can be done by means of transformations (1.14)–(1.18), that have not been used in our procedure yet. Once the fictitious components (1.25) and (1.28) are substituted into (1.14), we are brought to the formula which determines the true components E_x and B_x:

$$E_x = \frac{qx'}{(x'^2 + y'^2 + z'^2)^{\frac{3}{2}}}; \qquad B_x = 0. \qquad (1.29)$$

On substituting expressions (1.25) and (1.28) into transformations (1.15) and (1.18) we arrive at two simultaneous equations in two unknown components E_y and B_z:

$$E_y - \beta B_z = \frac{qy'}{\gamma(x'^2 + y'^2 + z'^2)^{3/2}}; \qquad \beta E_y - B_z = 0. \quad (1.30)$$

After solving this pair of equations and taking into account the second equation (1.10) (See page 47) we arrive at the following expressions for the components E_y and B_z:

$$E_y = \frac{q\gamma y'}{(x'^2 + y'^2 + z'^2)^{3/2}};$$

$$B_z = \frac{q\beta\gamma y'}{(x'^2 + y'^2 + z'^2)^{3/2}};$$
(1.31)

The expressions for the components E_z and B_y are obtained similarly:

$$E_z = \frac{q\gamma z'}{(x'^2 + y'^2 + z'^2)^{3/2}};$$

$$B_y = \frac{q\beta\gamma z'}{(x'^2 + y'^2 + z'^2)^{3/2}};$$
(1.32)

But the procedure of the charge's "move" to the real world is not finished yet. Though the left sides of the equations (1.29), (1.31), and (1.32) are the true field components, the right sides of these equations are still represented in terms of fictitious coordinates x', y', z'. To complete the "move", we have to change the primed variables for the true ones by using transformations (1.11) and (1.12). This will bring us to the following final expressions for the field components of a moving charge:

$$E_x = \frac{q\gamma(x-vt)}{[\gamma^2(x-vt)^2 + y^2 + z^2]^{3/2}}; \qquad B_x = 0;$$

$$E_y = \frac{q\gamma y}{[\gamma^2(x-vt)^2 + y^2 + z^2]^{3/2}}; \qquad B_z = \beta E_y; \quad (1.33)$$

$$E_z = \frac{q\gamma z}{[\gamma^2(x-vt)^2 + y^2 + z^2]^{3/2}}; \qquad B_y = -\beta E_z;$$

At last we have reached our destination. There are no primed variables here, and the real vectors **E** and **B** are expressed exclusively in terms of the real variables x, y, z, and t. Note that time t enters here only through a binomial $x-vt$, which is nothing else but a distance – along the x-axis – between the moving charge and the mov-

1.6. ELECTROMAGNETIC FIELD OF A CHARGE MOVING AT A CONSTANT VELOCITY

ing reference point at which the field is considered. The fact that time takes part in only such a combination means that the field is propagating together with the moving charge. To make it certain let us rewrite equations (1.33) for the coordinate

$$x = X + vt, \qquad (1.34)$$

which defines the reference points which are moving ahead of the charge at a distance X. A substitution of this coordinate into (1.33) leads to

$$E_x = \frac{q\gamma X}{[\gamma^2 X^2 + y^2 + z^2]^{3/2}}; \qquad B_x = 0;$$

$$E_y = \frac{q\gamma y}{[\gamma^2 X^2 + y^2 + z^2]^{3/2}}; \qquad B_z = \beta E_y; \qquad (1.35)$$

$$E_z = \frac{q\gamma z}{[\gamma^2 X^2 + y^2 + z^2]^{3/2}}; \qquad B_y = -\beta E_z;$$

We see that time t takes part in the equations only implicitly — through variable X. This means that at every point which accompanies the moving charge the field does not depend on time or, in other words, the field accompanies the moving charge. Since it behaves as if it were attached to the charge, it is usually called an intrinsic or associated field.

1.6.2. The Lorentz contraction of electromagnetic field

Let us see now how the motion of the charge affects its electromagnetic field. This can be done by comparing the field (1.35) of a moving charge with the field of the same stationary charge given by

$$E = \frac{q}{r^2}; \quad \mathbf{B} = 0. \qquad (1.36)$$

To make (1.36) comparable with (1.35), the fields **E** and **B** should be resolved into their components just in the same way as was done to the fictitious field **E'**, whose components were defined by (1.28). This will bring us to the following analog of (1.28) and (1.25):

$$E_x = \frac{qx}{(x^2+y^2+z^2)^{\frac{3}{2}}}; \quad E_y = \frac{qy}{(x^2+y^2+z^2)^{\frac{3}{2}}};$$

$$E_z = \frac{qz}{(x^2+y^2+z^2)^{\frac{3}{2}}}; \quad B_x = B_y = B_z = 0.$$

(1.37)

The fields of the charge in motion (1.35) and the fields of the charge at rest (1.37) are not only comparable but also convertible. When the moving charge stops, (1.35) turns into (1.37) as was to be expected. We can make sure of it by substituting respectively 0 and 1 for β and γ in (1.35).

Now we can figure out the details which make (1.35) different from (1.37). The first thing that catches the eye in (1.35) is the presence of the magnetic field of B_z and B_y. This is not surprising, because the moving charge is a current, and the current must have a magnetic field, whose lines have the shape of circles strung onto the trajectory of the charge. It's just the way they would look, had we plotted them according to (1.35).

Let us turn to the electric field. Comparing the electric field components in (1.35) with those in (1.37), we see that they have much in common, but there are two formal distinctions. The first of them concerns the coordinate X, which is multiplied by a factor of γ wherever it happens to appear. This regularity signifies that setting the charge in motion with constant velocity makes its field contracted along the motion by a factor of γ as shown in Fig.11(c). If for example $\gamma = 2$, then the multiplication of all the possible values of X by 2 means that the field retains its pattern provided every point of the pattern has become twice nearer to the charge along the x-direction. Such a shrinking of the field was discovered by Lorentz and is called the Lorentz contraction.

The second distinction caused by motion is also connected with the Lorentz contraction. It involves the transverse components E_y and E_z, which (as seen from (1.35)) have not only their denominators changed in accord with E_x, but also their numerators multiplied by γ. Thus, the Lorentz contraction of the electric field is always accompanied by an appropriate amplification of the transverse field components. This is inevitable because for the longitudinal shrinking of the field to be regular – without any distortions – the density of the electric lines looking forward and backward must decrease, while that of the lines looking sideways must increase, as can be seen in Fig.11(a) and (c).

1.6.3. Physical mechanisms of the field contraction

The Lorentz contraction of the field can be explained even without any mathematics – just from the law of electromagnetic induction. Because the moving charge is an electric current, it must have a magnetic field, whose lines of force look like rings strung onto the current, i.e. the trajectory of the charge. Most rings are situated nearer to the charge. Far ahead of the charge and far behind it, there are fewer rings, because the farther from the charge, the weaker the magnetic field. If the charge passes by a stationary observer, the latter can use an appropriate magnetic probe (say, a compass) to detect this field and to see how it gradually grows up as the charge is coming nearer, and then dies out as the charge gets far off. Hence, at an arbitrary stationary point of space, the magnetic field increases before the charge and decreases behind it. But according to the law of electromagnetic induction, this must cause the vortical electric field whose lines of force encircle the changing magnetic flux. This field is shown in Fig.11(b) as a difference between the field of the charge in motion and the field of the same charge at rest. In accordance with Lenz's law, it must be directed against the current, when the current is rising, and along the current, when it is falling. Therefore, the induced electric field is opposed to the motion before the charge and is aligned with the motion behind it. Its lines of force first converge on the charge both from in front of the charge and from behind it, then reflect from the charge sideways and close on themselves in some remote regions of space. (See their dashed continuations in Fig.11(b)). If the charge is of a point size, then the lines get closed in infinity.

Only a certain part of the entire field is shown in Fig.11(b) – the part that is caused by motion and is responsible for the Lorentz contraction. To turn it into the net field, we have to complete it with the field of the stationary charge shown in Fig.11(a). In other words, we have to add (or superimpose) the two fields shown in Fig.11(a)

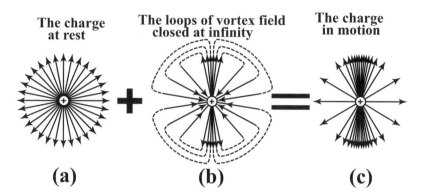

Fig.11. The origination of the Lorentz contraction of the field which is created by a positive point charge "+" moving with a constant speed $v = 0/94c$ in a horizontal direction. The patterns of the electric field, shown in figures (a) and (c), belong respectively to the charge at rest and in motion. The difference between them is shown in figure (b). According to the law of electromagnetic induction, this difference is generated by the charge's magnetic field, whose lines of force, for clarity, are not shown here. They have the shape of circles threaded onto the trajectory of the charge. Since the lines of force in figure (b) belong to a vortical electric field, they do not terminate or originate in the charge and form closed loops, whose behavior at infinity (i.e. far away out of the frame of the picture) is shown conventionally by dashed lines.

and (b) to arrive at the net field given in Fig.11(c). Comparing the pictures (a) and (c), we can see that under the action of motion, the field becomes weaker in front of the charge and behind it (there is fewer lines there), and stronger aside of the charge, where the lines are much denser. This is nothing else but the Lorentz contraction, whose strict description is given by equations (1.35).

When the speed of the charge approaches that of light, factor γ is unlimited in its growth and the lines of force get directed right sideways They can become so dense that actually form a flat disk with a very strong field within it. So, the field of a moving charge is always asymmetric. It looks different from the front and from aside.

Though the field of a moving charge experiences a γ-fold contraction, the compressed field somehow manages to save its pattern in the process of motion. If taken formally, this property seems quite

1.6. ELECTROMAGNETIC FIELD OF A CHARGE MOVING AT A CONSTANT VELOCITY

natural. There are no obvious reasons indeed for preventing the associated field from following the moving charge. Just like the clothes you are wearing travel together with you. But at the second thought, this phenomenon captures our imagination. Indeed, let us take a certain point M away from the charge which accompanies the charge at the same velocity **v**. If the charge suddenly came to rest or turned aside, the field at the point M would "learn" about it not instantly, but some time later, and until then, the field would continue to propagate with the same velocity **v**, as if nothing had happened. The field at M, though, "knows" where the charge was at a certain earlier moment of time, passing through some previous point of its trajectory. The information from there had time to arrive at point M, because this information propagates with a speed of light c, i.e. faster than the charge. That's why the associated field is propagating with the speed v, and not the speed c, which would be expected if it were a free field, propagating without any charges. But look at Fig.11(c). Put the point M anywhere you like. Where does the electric field stretch its lines from that point? It stretches them not to the place where the charge was some time before, but right to the place where it is expected to be now, though in fact it might never get there.

We know the free electromagnetic field to be able to propagate independently — without following any charge or current. Therefore the associated field has many chances to break away from the charge instead of following it obediently. It could outrun it, or dodge aside, or, while following the charge, it could change its pattern. It could even stretch its lines backwards so as to retard the charge and disturb the uniformity of its velocity. But nothing of the kind occurs. Looking at Fig.11 you may indulge yourself with comprehending anew one of the greatest laws of Nature — Newton's First Law — a body which is at rest or is moving with constant velocity continues to do so unless some external force is applied to it. There are lots of charges and fields inside every body, and if the associated fields did not follow their charges with such a loyalty, Newton's First Law would be violated.

Such behavior of the associated field seems surprising because the phenomenon of field propagation is rather sophisticated even in the case of a single charge. We can examine it in every detail by solving the Maxwell equations, say, with the help of a computer. Such computations will clarify the contributions of Coulomb's law, of Ampere's law, of Faraday's induction law, of the displacement currents. Every process in every local region of space for every moment of time will become absolutely plain and quite understandable. But in spite of it, the eventual result, obtained for the associated field as a whole, is still surprising and looks mysterious.

The associated field accompanies the charge with such an incredible loyalty only in the case of uniform velocity. This harmony is destroyed as soon as the velocity of the charge changes. What will happen indeed to the associated field if, for some reason, the charge suddenly increases its velocity? In accordance with the formulas (1.35), the associated field must contract. But can the field do it instantaneously? The answer is "no". If far from the charge, the field cannot "learn" immediately about the change in the speed of charge's motion. Some time must have passed before this information from the charge arrives at the reference point, and until then the old field pattern is retained there. The associated field will begin to re-establish its pattern first near the charge and only after that in more distant regions of space. A complicated phenomenon of field re-establishment is started. A portion of the field, can break away and begin its own life independent from the charge. Radiation will take place. And whatever happens then to the charge, the runaway field will never learn anything about it. As soon as the radiation ceases, the remaining field will acquire the pattern determined by the expressions (1.35) based on the new value of the charge's speed. The process of emitting the radiation has come to an end. Later on the new associated field will follow the charge, conserving its new pattern just in the same way. If the velocity of the charge is changing continuously, i.e. if it is accelerating or retarding, then the emission of radiation persists continuously.

Similar phenomena may take place even in the case of the charge moving with a constant velocity, provided some external objects (say, some conducting bodies) happen to be near its trajectory. Then the associated field, determined initially by expressions (1.35), will have to re-establish its structure, adapting to those objects, and there will be a radiation. There are many various causes of the emission of radiation. But all of them involve some disturbance of the associated field and its subsequent re-establishment. So it can be assumed that it is the disturbance of the associated field of a charge or of a current that is responsible for any kind of electromagnetic radiation.

1.7. What Einstein did to the Wonderland of Lorentz

where we try to show how Einstein changed the physical interpretation of the Lorentz transformations and how Einstein's revolution looked like in the eyes of the contemporaries

1.7.1. Wonderland of Lorentz

While dealing with the field of a charge moving with a uniform velocity, we saw how effective the Lorentz transformations were. It was due to them that we succeeded in getting the field of a moving charge, not even knowing how Maxwell's equations look like. The procedure of obtaining the desired field was rather simple. A point charge q, moving in our real world of non-primed variables and having some true coordinates x, y, z, was moved to the imaginary land of Lorentz and placed there at a certain fixed place. At first this charge had no electromagnetic field – it was naked. But since the charge was not in motion there, it was easy to get it dressed in an electromagnetic field and then return it back to our real world together with its new attire. In our real world the charge started to move again with its field being contracted in the direction of motion as shown in Fig.11(c).

The Lorentz transformations (1.11)–(1.20) (see page 48) served as a tool for resettling the charge forth and back. It makes sense to review them once again. It is they that are to be responsible for the properties of Lorentz's imaginary world, which so far has been used as a very comfortable cloak-room for dressing imaginary charges in their imaginary electromagnetic attire. Let us begin with the transformation (1.11). It looks the simplest of them and comprises addend vt and factor γ. The meaning of vt is quite clear. It is a "brake" that instantly stops the charge or the current as soon as it is resettled to the land of Lorentz. It is this term that is responsible for the global rest of charges and currents in that land. The factor γ works in a different way: It stretches all the bodies along the x-axis making them longer by a factor of γ. The transverse sizes of the bodies are retained in accordance with the transformations (1.12). The stretching of solid bodies is not surprising as it occurs in an imaginary world.

The next transformation (1.13) deals with the time t' that, for some reason, proved dependent not only on the true time t, but also on the spatial coordinate x. What could it mean? To answer this question, let us take a timepiece in our real world and send it traveling along the x-axis with a speed v. We wonder what will happen to it if it is resettled to the land of Lorentz. The substitution of $x = vt$ into the transformation (1.11) shows that the timepiece will be at rest there ($x' = 0$). On the other hand, the same substitution into (1.13) results in $t' = \gamma t(1 - v^2/c^2) = t/\gamma$ and tells us that the fictitious clock in the world of Lorentz ticks γ times more slowly than its real copy in our real world. Let us now look at the time t' from another side. Suppose that a continuous array of timepieces is moving in our real world along the x-axis in a single file, all of them showing now the same time $t = 0$. Where are now the images of these clocks in the world of Lorentz, and what time do they show there? To answer these two questions, it is enough to substitute $t = 0$ into the transformations (1.11) and (1.13) and then to see what x' and t' will be equal to. The substitution into (1.11) gives $x' = x\gamma$. This means that the clocks are distributed γ times wider along the x'-axis in the world of Lorentz than in our world. On the other hand, it follows from the equation (1.13) that

$$t' = -\frac{\gamma v x}{c^2} = -\frac{v x'}{c^2}.$$

Thus, the readings of the clocks depend on their location on the x'-axis — the farther along the x'-axis, the shorter the time shown by the clock. Moving along the x'-axis is like flying from Vladivostok to Moscow; you are resetting back your watch all the time. The world of Lorentz proved to be divided into time zones, gradually changing each other. It turned out that time and space are coupled together there. This was a good reason for Lorentz to introduce the concept of a local time, to which he, certainly, did not ascribe any physical meaning as it belonged in an imaginary world.

Let us continue our excursion. We will proceed at once to the transformation (1.19), that looks rather odd. Imagine that there is a current-carrying rectangular frame which lies in a vertical plane and moves horizontally in this plane with a speed v. The velocity of electrons with respect to the frame will be denoted by w. The following relations will then be valid in the upper and lower conductors of the frame: $u_x = v \pm w$. A substitution of these two values into the transformation (1.19), leads to

$$\rho' = \rho\gamma\left[1 - \beta^2 \mp \frac{vw}{c^2}\right]. \tag{1.38}$$

We see that the density ρ' of the charge is distributed between the conductors of the frame in a non-uniform way – the upper conductor contains less charge than the lower one. After being resettled to the world of Lorentz, our frame stopped moving, but for some reason certain electrons have moved from the upper conductor into the lower one. In other words, the frame has acquired an electrical polarization. This thing would be very surprising if it happened in our real world.

It's time now to concede that we have been unjust calling the imaginary land of Lorentz "a cloak-room". It is not a miserable cloak-room for dressing the charges in their electromagnetic attire, but rather a Wonderland built up by the power of human imagination. Lorentz enjoyed taking mental walks about his fairy possessions, accompanied sometimes by his like-minded disciples. The most famous among them was the mathematician and philosopher Poincare. Aware of the unreality of that world as they were, they still could not take their eyes off these landscapes, where the most strange and unusual phenomena were so perfectly matched. Lorentz even lodged an imaginary observer there and arrived at the conclusion that the latter would never learn anything about the wonders occurring in his world. It turned out that all those wonders could be noticed only from our world and were invisible from inside.

1.7.2. Evacuation of the first miracle

On went life and once upon a time a curious thing happened to the Wonderland of Lorentz. Just like Pinocchio cut out of a log by uncle Jepetto, the world of Lorentz all of a sudden squeaked to its creator in a very shy though a rather distinct voice: "I do exist!" At the moment, Lorentz did not suspect yet that it was the first shock of a coming earthquake. He did not know that that earthquake would cause a tremendous shift in ideas and that all the miracles of his imaginary wonderland would eventually leak out to the real world surrounding us.

The events developed like this. Trying to explain Michelson's experiment, which was a sensation at that time (we will speak of it in the next section), Lorentz took on the following task. There is a certain stationary system of electric charges that are in equilibrium with each other (This system stood for a solid body). The coordinates x_0, y_0, z_0 of every charge of this system are taken as given values. All the elements of the system interact only through the electric and magnetic fields. He was looking for the answer to the following question: If this system of charges is set in motion with a known constant velocity, what changes, if any, are expected? To all appearances, the motion had to affect the mutual location of the charges, so that the coordinates x, y, z of every charge will differ from the initial values x_0, y_0, z_0 not only by the mere addend vt. We know the field of every separate charge to be contracted by a factor of γ. Maybe a system of charges behaves in a similar way. But it was not a guess that could satisfy Lorentz. What he wanted was a derivable result. However, the round-trip of charges to the Wonderland and back could be made in only one direction – from the Wonderland to our real world. The resettling of the equilibrium system of charges to the Wonderland was out of the question, because the size of the system in our real world was unknown in contrast to the case of a point charge, whose size had been equal to zero everywhere and posed no problem.

To get out of this deadlock situation, Lorentz had to endow the Wonderland with an additional property that did not follow from the transformations (1.11)–(1.20). It had to be supposed that the imaginary charges in the imaginary world interact with each other as if they were quite real stationary charges. In other words, there appeared forces in the imaginary world (to be followed by the accelerations and masses) as if the imaginary world were real. "Our imaginary world is borrowing more and more features from the real world" – something like that was stated by Lorentz in his works published at the time.

Lorentz knew how imaginary (primed) forces are connected to the true (non-primed) ones that act in our real world. This connection was realized through the transformations (1.14)–(1.18) for the true and imaginary electromagnetic fields. Once these fields are known, formula (1.6) (see page 25) can be used to get the desired forces. Finally, the relationship between the true and the imaginary forces proved to be very simple. Here it is (the derivation will be given in Section 2.1):

1.7. WHAT EINSTEIN DID TO THE WONDERLAND OF LORENTZ

$$F_x' = F_x; \qquad F_y' = \gamma F_y; \qquad F_z' = \gamma F_z; \qquad (1.39)$$

Thus, when resettled from the real world into the imaginary one, the component of the force along the x-axis (F_x) is transformed into itself while the two other components (F_y and F_z) become stronger by a factor of γ. Only one very simple corollary of (1.39) was used by Lorentz in his derivation: If the forces in the imaginary world are equal to zero, i.e. the particles of a body are in equilibrium with each other, then in our real world they also vanish. When a solid is resettled from the wonderland into our real world or vice versa, the equilibrium of its particles is retained. The solid body remains as solid as before the resettling.

The way now is paved to the estimation of the effect of motion on the size and shape of a solid body. To make this estimation, we needn't resettle the body from the real world into the imaginary one. That would be hopeless because we don't know the size and the shape of the moving body subject to this resettling. But what if we will do it in the opposite way? What if we let our body be born in the Wonderland of Lorentz instead of being brought there? All the moving charges and currents are at rest there, aren't they? Isn't it just the right place for our system, whose size is given precisely for the case of its being at rest. Assume it has been born in the imaginary world with its size and shape answering the initial conditions of the problem. Then it will be in equilibrium there, and when resettled to our world, it will turn into a moving body (this time the brake vt will act as an accelerator). During that resettling the shape and size of the body will undergo some change, but the equilibrium will be retained, transformations (1.39) providing it.

How after all, will the dimensions and the shape of the body change when it is resettled from the world of Lorentz to the real world? It is not difficult at all to answer this question. If, while being resettled "there", the body stretches in the direction of motion, then on the way back everything will be in the opposite way. The body will contract by a factor of γ in the direction of motion, keeping its transverse sizes unchanged. That was the conclusion, made by Lorentz whose reasoning has been reproduced here in a somewhat simplified form.[1]

The longitudinal contraction of the moving bodies was a great discovery, which Lorentz unpretentiously called "a hypothesis". He just wasn't sure in the electromagnetic origin of the molecular forces. He could only suppose that these forces behaved in the same way as electromagnetic interactions. This discovery meant that the world of Lorentz has lost one of its miracles. Before Lorentz formulated his hypothesis, everyone thought that motion did not affect the size of bodies. If that had been true, then, after resettling to the World of Lorentz, the body would stretch by a factor of γ acquiring an unusual form. In reality, however, this process goes in the opposite way: It is our world where the moving body has an unusual, contracted form. And when resettled to the World of Lorentz, the body stretches and becomes a very ordinary solid body at rest, i.e. it takes its natural shape – neither contracted, nor stretched.

The wonderland lost one of its miracles, and our world gained the first miracle of the group of phenomena that later would be named "relativistic effects". As for the Land of Lorentz, it might be called now not only a cloak-room for dressing the charges in their electromagnetic attire, but also an obstetric ward: When considering complicated systems of charges, we needn't resettle them first to the world of Lorentz and then back. They should be born right there, and after dressing them in their electromagnetic attire they should be immediately resettled to the real world. Having lost one of its miracles, the World of Lorentz did not suffer much from that and still looked almost as mysterious and fantastic as before. The lost miracle concerned only the size of a solid body and did not belong to the most striking wonders of Lorentz's gallery.

1.7.3. Einstein takes the floor

Should Lorentz continue taking on and solving other similar problems concerning the effect of motion on various stable equilibrium systems (such as clocks, current-carrying loops and so on), then each solution of such a problem would be followed by resettling of

---

1 In fact, the reasoning of Lorentz was much more refined and complicated. At the time when Lorentz succeeded in predicting the contraction of the moving bodies, his transformations were still under development. They were eventually formulated considerably later. We do not go into such historical details so as to avoid distraction from the essence of the issue.

the next miracle from the Wonderland to the real world, so that eventually the imaginary world would lose all its miracles. It is hard to say how many years it would take to complete this program, and whether Lorentz could have accomplished it by himself. This is a rhetorical question because, in 1905, Einstein postulated (i.e., stated without any proof) the assertions that turned upside down not only the Wonderland of Lorentz, but also all the ideas of physicists about the world surrounding us.

Einstein's ideas might sound in the minds of Lorentz and his disciples as the following three assertions.

1. Why do you regard your imaginary world as wonderful? It is our real world that is wonderful indeed, and not the imaginary world of yours. As for the imaginary world, it is not wonderful at all. It is just an ordinary cloak-room where charges are dressed in their electromagnetic attire, and an obstetric ward where no miracles may be born by motion, because all the charges and currents (in your equilibrium models) are stationary there.[1] Weren't it you who invented that world in order to solve Maxwell's equations for stationary bodies instead of doing it for the same bodies in motion? While your world is only occupied with stationary equilibrium systems that are quite ordinary, our real world is populated by not only stationary but also moving systems that behave in most miraculous way in full accord with your transformations. Solid bodies reduce their length in the direction of motion. You made sure of it for yourself. But this is only one of the miracles. When in motion, the clocks are ticking more slowly. If there are several clocks moving in a single file and synchronized with each other through whatever you like (for instance through electromagnetic signals), they will show different times. If the clocks were synchronized through sound signals, the result would be the same. And even if you nailed the hands of the clocks to a common solid joist, it would not help. While setting from rest to motion, the joist would bend queerly and the clocks

---

1 Einstein was a very tactful man and he never used, of course, such expressions. The author only makes an attempt to reproduce here how Einstein's proposals might be interpreted by the confused representatives of the pre-relativistic physics. Some figurativeness of expressions is destined here to facilitate the comprehension of new and unexpected ideas.

in different places would show different times. Different time zones will appear in a moving laboratory (those very time zones were erroneously ascribed by you to the imaginary world; isn't it you who has coined the term "local time"?), and when going across these time zones, the hands of the clocks will shift back or forth just by themselves. When set in motion, an electrically neutral current-carrying loop will acquire an electric polarization that gives rise to an electric field around it. Indeed, all these miracles occur in our real world in full accordance with the well-known laws of nature. If we have not noticed them before, that was either because most (but not all) of them become noticeable only at a speed close to that of light, or else because of our innocence in interpreting the laws of nature, discovered long ago. As for me, I deduce all those miracles not from particular laws of nature, but from the fundamental idea which I call the principle of relativity and which extends beyond the limits of your imaginary world as well as beyond the whole electrodynamics. It covers not only the theories available nowadays, but also those which will be developed in the future.

We try here to reproduce how the first group of Einstein's ideas might be perceived by his contemporaries. It is impossible to overestimate the importance of these ideas for the evolution of physics. But Einstein declared much more. His next statement was no less striking than the first:

2. So, you agree that your imaginary fairy world of primed variables is not fairy at all. It is quite an ordinary world in which the equilibrium systems are always at rest. Now I would like to tell you that this world is not imaginary either. It is a real world of a real observer who, with a set of measuring instruments, accompanies the system of bodies involved. According to the instruments at rest, the bodies in motion undergo a lot of wonderful transformations. But the same transformations are experienced by the moving instruments and standards. Any measurement is a comparison with a standard. But the result of the measurement will not change at all if both the object of measurement and the standard undergo the same conversions, no matter how wonderful those conversions might be. That's why the moving observer, accompanied by the set of measuring instruments, will regard his "wonderful" moving system of bodies as an ordinary system at rest. The results of his measurements will be the same as those of an observer at rest, whose objects of measurement are also at rest. However sophisticated the measurements of the moving objects by the moving observer might be, whatever advanced his experiments might be, he will never be able to establish the fact of his motion unless he extends his activities beyond his moving environment and overhears what the instruments at rest are "talking" about his measurements.

Having converted the world of Lorentz from imaginary into real, Einstein endowed Lorentz's primed variables with a new physical meaning. By doing so, he has changed the physical sense of the transformations themselves, because in physics any regularity is determined not only by its mathematical formulation, but also by the meaning (i.e. by the physical definitions) of the variables participating in that formulation. Compare, for example, the formula which determines the force of interaction in Coulomb's law with that in the law of universal gravitation. If the constants of proportionality in both laws are reduced to unity by a relevant choice of the units of measurement, the mathematical expressions for the two laws are identical. Does it mean that the laws themselves are identical too? Of course not! It is the charges that interact in the first of them, and masses that interact in the second law, which makes a lot of difference. The same refers to the Lorentz transformations. When taken in Lorentz's or Einstein's interpretations, they have different physical meaning in spite of identical mathematical representation. The *non*-primed variables have the same physical meaning in both interpretations. They specify the results of the measurements, made in the moving system of bodies by instruments at rest. As for the primed variables, they were interpreted by Lorentz and Einstein in different way. Lorentz used them as nothing else but the auxiliary mathematical quantities which simplify the procedure of solving Maxwell's equations. As for Einstein, he declared them to be the results of the measurements which are made in a moving system of bodies by *proper* instruments, i.e. by the instruments that accompany their objects of measurement. In other words, Einstein has "revived" the primed variables by assigning them to the real world of a moving observer, which made the imaginary world of Lorentz just unnecessary. Though nowadays the Lorentz transformations are used much wider than in the pre-Einsteinian times, no one says now: "Let us move over to the imaginary world of Lorentz." Instead we say: "Let us switch over to the moving (or proper) frame of reference." As for the virtual, i.e. non-primed variables, they are associated nowadays with the frame of reference at rest (called usually a laboratory frame). Nowadays physicists routinely pass from one frame of reference to another by means of the Lorentz transformations, looking for the frame which is most convenient for solving the problem involved.

3. Einstein's third statement established the complete symmetry between the measurements of the observer at rest and those of the observer in motion. If two identical standard rods – one at rest and the other in motion – are compared with each other, then according to the observer at rest, the moving rod is shorter than the one at rest, and at the same time, according to the observer in motion, it is the rod at rest that is shorter than the one in motion. At first sight, such an uncompromisable symmetry might seem rather paradoxical. It was only Einstein who dared to predict this result. It does not depend on the observers, of course, but on the sets of instruments used by them. In the first case the instruments are at rest, while in the second case they are in motion, having acquired new properties. But do we really need a "set" of instruments here? Isn't one standard rod (either at rest or in motion) sufficient to measure the length of the other rod – an object of the measurement (either in motion or at rest)? It turns out that it is necessary to have also at least two clocks. If the initial (zero) marks of the two rods coincide at a certain moment of time, then their end marks must be compared at the very same moment of time. To do so, one needs two clocks attached to the initial and end marks of the measuring rod. If the standard rod and the clocks move together, the motion affects the readings of the clocks, that fall within different time zones. The motion shortens the rod, but it also changes the readings of the clocks. The latter effect is stronger than the former, and the moving rod will eventually be measured as being longer than the stationary one. This symmetry between the observer at rest and the observer in motion (or to put it more correctly, between the instruments at rest and those in motion) extends to all the other kinds of measurement.

But if the observers at rest and in motion are equal in their rights, then it is impossible to establish which one of them is at rest and which one is in motion. The need in Lorentz's ether disappears because any motion relative to the ether turns out to be undetectable. This means that the principle of relativity of the motion with uniform velocity, formulated at the time of Galileo, is valid not only for mechanical, but also for electromagnetic – and, in general, for any other natural phenomena – known to us today. But this principle works in a much more complicated way than it was believed at the time of Galileo. At that time scientists thought that the motion with a constant velocity does not affect any properties of a moving body, and therefore that motion may be only relative. However, the development of electrodynamics in pre-Einsteinian times showed that the motion with a uniform velocity does affect many physical phenomena, when these phenomena are estimated by instruments at rest. That led to the idea of the ether. Then Einstein comes and states that, *due to the fact* that motion affects the properties of bodies, it is impossible to detect the ether as well as the motion with uniform velocity. Perhaps you feel how strange this statement sounds. Its second and first parts seem to contradict each other. Should Einstein have said: "In spite of the effect of the motion...", that would sound much more logical than "due to..." But the thing is that the relativity of the motion with constant velocity holds in unusually unique and quite definite way, discovered by Einstein – the different phenomena caused by the uniform velocity of motion act in concord, and, as soon as the moving observer gets a chance to detect his own motion through the ether by means of a certain physical effect, there appears another effect, seemingly irrelevant, which cancels it altogether. The coordination of these two effects is performed by the Lorentz transformations that act like a conductor of a good orchestra. As Einstein showed, those transformations can be obtained not only from classical electrodynamics, but also in a much more general way. In order to get them, Einstein solved the following problem: "how must the motion with constant velocity affect the length of the moving bodies and the readings of the moving clocks to make it impossible to detect the absolute motion?"

That's why Einstein did not present his considerations in the order used here, i.e. from electrodynamics and mechanics to the principle of relativity. He did it in the opposite way. He began by borrowing only one rather general idea from Maxwell's electrodynamics: the speed of light does not depend on the motion of the source. It was the second postulate of special relativity. If taken alone, it could be comprehended without problems. It might even seem that an ether had come into being for a moment. If the speed of light did not depend on the motion of the source, it would be quite natural that light propagates through the ether just like sound through the air. But Einstein gave up this idea immediately by advancing his first postulate: changes in the states of physical systems obey the same laws in all the frames of reference moving at a uniform velocity. In other words, it is impossible to detect the absolute uniform velocity of motion and, hence, there cannot be any ether in nature. To conciliate the two postulates, Einstein, first, admits that motion changes the properties of bodies (the ether has turned up for a moment once again as a probable material cause for this influence), and, secondly, he postulates this influence to come about not in an arbitrary way but on the certain special condition of excluding any possibility of detecting the absolute motion (if nature does answer this requirement, then this time Lorentz's ether will vanish for long). First of all, in order to satisfy his first postulate, Einstein elucidates the effect of motion on the length of rods and on the tick of clocks. Suppose the speed of light is measured at first by stationary instruments, and then by the instruments moving with a speed v along the ray of light. It is evident that in the first case the measured value is c. In the second case, the measured value would be $c-v$ unless the instruments were affected by the motion. But the real instruments do in fact be affected by the motion making the measured value different from $c-v$. What is this value? Einstein assumed that it was c, because otherwise his first postulate would be violated. To satisfy this requirement, the readings of the moving instruments ("distorted" by motion) must correspond to the readings of the stationary instruments (not "distorted" by motion) in quite a definite way, described by certain algebraic relations. Einstein has obtained these relations, which turned out to coincide with the Lorentz transformations (1.11)–(1.20). In other words, the moving rod must be γ times shorter than the same rod at rest (as was shown before by Lorentz), the tick of a moving clock must be γ times slower than that of the same clock at rest, and the hands of a moving clock must reset themselves as soon as the clock changes its place relative to the other moving bodies (no one could even suspect this in the pre-Einsteinian epoch). This signified that space and time display properties previously unknown and that they are coupled together. As for electrodynamics, this meant that the next two miracles (the clock tick slowdown and the local time) had left the fairy world of Lorentz for the real world of ours.

But Einstein's activities did not end there. He required that Newton's laws must also obey his first postulate. It might seem that he was forcing an open door. No one doubted that the laws of mechanics satisfied the Galileo principle of relativity and that, in mechanics, the uniform velocity made sense only relative to other bodies. Were it not for electrodynamics with its ether, Einstein would have nothing to do there. This reasoning held until Einstein discovered the new properties of space and time. After that, Newton's laws acquired a different form for the laboratory at rest and that in motion. The ether was about to reappear – this time in mechanics. But Einstein coped with this problem as well. To do it, he did not even have to alter anything in or add anything to Newton's laws. It was sufficient to assume that the mass of bodies was proportional to γ, i.e. that it depended on the speed of their motion. Strictly speaking, this idea was not new. Long before Lorentz and his disciples established namely that dependence of the mass of the electron on the speed of its motion through the ether. Now Einstein demanded that not only an electron in motion should be more massive than that at rest, but also that an electron at rest should be more massive than that in motion. The cause for such an "incredible" behavior was the same as before. The mass had to be measured by something, didn't it? Because the moving instruments possessed new properties, their readings were opposite to those of the instruments at rest. The effect of motion upon the mass of bodies was declared by Einstein as a universal property of all the bodies in nature. That led him to a "by-the-way" discovery of the mass-energy equivalence, that would prove to be of crucial importance for the evolution of physics.

1.7.4. *A miraculous survival*

Thus, the fairy world of Lorentz became a reality. It was done away with the ether. Mechanics and electrodynamics conformed obediently to Einstein's radical demands. But rather unexpectedly, both of them somehow managed to survive, having accommodated themselves to Einstein's approach. Since then, the scientists began to take into account the effect of motion on the length and mass of bodies, on the tick of clocks or their relative readings, and on many other physical quantities. But the laws themselves remained unchanged, though dressed up in new four-dimensional mathematical attire (with time as the fourth dimension). The classical physics that had been reigning in the pre-Einsteinian epoch remained formally as valid as ever. It was only the ether and the absolute space (introduced by Newton) that proved to be excessive (at least with regard to the motion by inertia), and therefore they were expelled from the theory.

The survival of mechanics and electrodynamics was not a matter of chance. It turned out that all relativistic discoveries (except the new approach to gravitation) were hidden in classical physics. Newton, Maxwell and Lorentz succeeded so much in their formulations of the fundamental laws of nature that the relativistic effects turned out to be hidden there. But the founders themselves did not even suspect it and therefore extended such notions as "the absolute space" and "the ether" to the motion by inertia

Had Einstein not developed relativity in 1905, it would have been discovered later on. But how and when? It might be that someone else would follow exactly Einstein's path. However, this seems hardly probable. It seems that the geniuses like Einstein are born only once in a few centuries. It looks more probable that if Einstein had not been born, relativity would be developed gradually, step by step by joint efforts of many scientists. Supported by Newtonian mechanics and the Maxwell-Lorentz electrodynamics, they would gradually extract from them all the relativistic effects that were deduced by Einstein through his brilliant guesswork, supported by his perfect knowledge of physics, and fantastic intuition. Wasn't it Lorentz who had managed to transfer his first "miracle" from the imaginary world into the real one? He was stimulated, though, by the Michelson experiment. The next "miracle" – the slowdown in clock tick – was demonstrated experimentally in a man-made experiment by Ives and Stillwell as late as in 1938. Perhaps theoreticians wouldn't have been waiting for so long. Investigating, at least mentally, different electrodynamical situations, they would have inevitably discovered that motion through the ether makes electromagnetic processes slower, which would have brought them to the reality of the imaginary time t'. After that the reality of the other primed variables would have been established just by analogy, so that all miracles of Lorentz would have been materialized. Then someone else would have presented the relativistic interpretation of the symmetry of the Lorentz transformations, the ether would have become undetectable and relativity would have triumphed at all events. But how much time would it have taken? It's hard to say, but perhaps several decades or so. But the deeper you comprehend the

1.7. What Einstein did to the Wonderland of Lorentz

roots of relativity and the simpler and more elegant it seems to you, the longer is the period you are likely to suggest.[1]

Dear Reader: If this last section turned out too complicated for your comprehension, don't despair and go on fearlessly reading the book. All the same, in the next section we will begin to explain everything once again in a much slower tempo. Though sometimes we will be recalling separate fragments of this section, we will never use them as a base for our further considerations up to Section 2.7, where we shall arrive once again at the same conclusions but on a much more detailed and well-grounded basis. The only thing that will be really needed is the electromagnetic field of a moving charge, discussed in Section 1.6. It is the contraction of that field in the direction of motion that will be a starting point of all the further conclusions. If you have not caught on the physical cause of that contraction, it would be better to reread Section 1.6 and maybe even some preceding sections. But bear in mind that it is not the mathematical formulations, but rather the physical reasons for the contraction that are really important there. It is necessary to grasp how the induction law works there, while the fact that the field contraction goes on not anyhow but exactly by a factor of γ may be believed even without proof: so many people after Lorentz have reproduced that derivation and used it in practice.

Einstein has developed his relativity in the most straightforward way, taking his two postulates as a starting point for all the derivations. He did not go scrupulously into particular physical reasons for relativistic effects. Such an approach was in accord with Einstein's endeavor to build his theory for nature as a whole, including not only the known phenomena but also those that had not been yet discovered. Taking his way, Einstein reasoned like this: if, according to the principle of relativity, something strange must occur, this "some-

---

[1] When many years later Einstein was asked about the contribution of the Michelson experiment into the development of Relativity, he answered that he did not remember if he had known about that experiment at all, but even if he had, it was not of crucial importance, because he had been always sure that the motion with constant velocity is always relative not only in mechanics but in electrodynamics as well. The unbiased character of this statement is confirmed by the fact that Einstein developed his relativity so promptly. It could have been done that way only by a deep inner inducement, and not under the pressure from some external circumstances.

thing" will be sure to occur, there always being good reasons to make it happen. Therefore he was mainly involved in the general conciliation of those seemingly irreconcilable contradictions rather than in the particular physical reasons for explaining them. It is that conciliation that should be regarded as Einstein's main achievement which paved the way to relativity.

But we are not einsteins – you and I. We are interested in details. We want to know exactly what it is that makes every rod shorter and every clock slow. So we will digest the relativistic effects gradually, basing on mechanics or electrodynamics and advancing by the way science might have taken if it had not been for Einstein. Step by step we will "revive" the Lorentz transformations one after another. Each time we will clarify the physical meaning of the relevant primed variable and dig out as best as we can the phenomena underlying every transformation. After this work is completed, the admirable world of Einstein will open before us. And the most striking thing to be grasped there is that that world is not invented or pictured. It is that very real world which we all live in.

PART 2

SPACE, TIME, AND RELATIVITY OF MOTION

2.1. Contracting rods

where we explain what makes a solid body contract after being set in motion with a constant velocity through the ether and why a moving observer fails to notice that contraction

2.1.1. Inside of a solid

Every solid body has a complicated internal structure. It consists of atoms with their positively charged nuclei and negatively charged electrons. It is hard to believe that the parts of such a complex system stay together instead of falling apart, and we have to apply ourselves to chop a log or to saw a steel bar. This is because the parts of the body are held by electromagnetic forces. We know that the charged particles attract or repel each other. There are also internal electric currents inside the solid. An electron rotating around a nucleus is an example of such a current. Such currents also attract or repel each other. Eventually every particle of a solid finds such a place for itself where the forces of attraction and repulsion are in balance with each other, so the average net force of interaction is zero. We have to use the word "average" here, because every particle of the solid may experience thermal oscillations; so it is a certain mean value of force that ultimately vanishes. A solid keeps its integrity because every particle manages to find such a position of equilibrium for itself. It is very important for the equilibrium to be stable. Whenever a particle is shifted from its equilibrium position, there arises a force that returns it back.

The shape, size, and density of a solid body are determined by the equilibrium positions of its particles. If there was a wand that could change these positions, then the shape and size of the body would change accordingly. For some materials, the role of such a wand is played by an electric field. Once a body made of such a material is placed between two oppositely charged plates, it begins to contract in one direction and extend in the other. Such substances are known as piezoelectrics.

In almost all bodies, the equilibrium positions of their particles depend on the amplitude of thermal oscillations or, in other words, on the temperature of the body. That's why solids expand or contract with the temperature. This phenomenon strongly depends on the material. For some substances (e.g. quartz) it is negligible.

And still there is a "wand" that can change the size and shape of any solid body regardless of its composition or structure. It is a motion of the body with a constant velocity. At first sight, this statement seems rather strange. It would be natural if we spoke of an accelerated motion – say, of a glass which falls down and breaks up. Then the change of the shape and size of the body would be on the face of it. The cause of the change would be also clear. It would be the abrupt deceleration of the body when getting in contact with the hard surface of the floor. If we accelerated the glass abruptly (for instance by inserting it into a gun tube and firing the gun), the glass would not stand it either, and would be destroyed. In both cases the destruction of the glass is caused by its acceleration, which is positive in the first case and negative in the second. A rubber ball, if dropped onto the floor, is not destroyed but flattens for a while in the process of its deceleration and restores its former shape after the subsequent acceleration is over. All these effects are caused by the non-uniformity of the velocity of motion. If the velocity is uniform, then the size and shape of the body seem independent of the speed of motion. When we travel by train, we notice the motion only due to the jerks of the car on the joints of the rails – only at the moments when the velocity of motion is non-uniform. The velocity of a plane is more uniform than that of a train. That's why we hardly feel any motion when on board of a plane. The velocity of a spaceship with its engine switched off is so uniform that even the most sensitive instruments fail to detect the motion. When the engine is turned on, the ship is accelerating, and everything inside the ship acquires some weight which can be registered by measuring instruments. But once the engine is turned off, the state of the ship becomes indistinguishable from the state of rest, no matter how large the velocity of the ship might be. This is in accordance with Galileo's principle of relativity that rejects an absolute motion of a single body as meaningless, and regards a relative motion of different bodies as the only sensible kind of motion. Therefore the size and shape of a solid body seem independent of whether it is at rest or is moving with a constant velocity. But once we look into the bowels of a solid body, our thoughts start flowing in the opposite direction.

We begin to realize that there is a lot of electric charges inside the body. When a body is moving at constant velocity, the internal charges also take part in the common motion and transform into electric currents that are either parallel or anti-parallel with the direction of motion. According to the laws of electricity, these currents must interact with each other. That alone is enough to cause some displacements in the equilibrium positions of different particles inside of the solid. And not only that. In addition to the currents, there arise vortical electric fields. These fields do not extend, though, beyond the limits of the body because the body as a whole is electrically neutral. But inside the body, its electric and magnetic fields are very strong and highly non-uniform – the closer to the charges, the stronger the fields. This non-uniformity could be detected by some imaginary microscopic field-strength probe stationed in the way of the moving body. According to the readings of such a probe, the internal fields would be time-dependent, with alternating regions of their weakening and growth. But the previous sections tell us that, according to the laws of electrodynamics, the variation in the magnetic field gives rise to an electric field, and vice versa. In brief, even when motion occurs at a constant velocity, the internal fields of a solid body are greatly different from those of the same body at rest. Hence the particles may have good reasons for changing their equilibrium positions, which would lead to a deformation of the body, whose size may be dependent on the uniform velocity of motion.

2.1.2. Length contraction

But how exactly is the size of a solid body affected by the motion at a constant velocity? Section 1.6 tells us that a single charge, if set in motion, has its electromagnetic field γ times contracted in the direction of motion. And so does every charged particle inside the body. If the distances between the particles did not contract accordingly, then the longitudinal equilibrium inside the body would be violated. Look indeed at the formula (1.35) for the component E_x. How would this component change if the x-coordinate remained the same while γ was, say, doubled? It's hard to answer because γ enters not only into the numerator but also into the denom-

2.1. CONTRACTING RODS

inator of the expression for E_x. It is much easier to answer another question: How will E_x change if the coordinate x reduces by a factor of γ in the process of the charge acceleration? It will not change at all. This follows indeed from (1.35). When the body is at rest, γ is equal to unity. In the process of the acceleration, γ grows while x decreases accordingly so as the product γx remains unchanged. But γ and x appear in the expression (1.35) for E_x only in terms of the product γx. Therefore E_x remains unchanged too. Thus, for the longitudinal equilibrium in a solid body to retain, all of the longitudinal distances between the particles must reduce by a factor of γ. Then the length of the body in the direction of motion will reduce in the same proportion.

Now let us take a look at the transverse forces inside the solid. How will they change when all the longitudinal distances become shorter by a factor of γ while the transverse distances remain the same? Formula (1.35) gives an answer to this question again. Imagine two positive charges q that are flying parallel to each other with the same constant velocity v, being separated by a distance y. These charges are shown as a pair of circles $\oplus$ in Fig.12. How will their repulsive force change when they are set in motion? From the one hand, this charges will repel each other under the action of their electric fields, and from the other hand, they will attract each other through their magnetic fields — because they are currents of the same direction, which attract each other in accordance with Ampere's law. When the charges were at rest (Fig.12(a)), they were not attracted at all, while the repulsive force was equal to qE_0 where E_0 was the electric field of a stationary charge at a distance y from the charge:

$$E_0 = \frac{q}{y^2}.$$

Now let us look again at the expressions (1.35) for the fields of a moving charge. Let us place one of our two charges at the origin of the coordinate system, whose y-axis extends between the charges as shown in Fig.12. How will E_y change if the charges are set in motion along the x-axis while their separation y remains the same? By putting $x = z = 0$ (the origin of the frame coincides with the charge that creates the field), we get

$$E_y = \frac{\gamma q}{y^2} = \gamma E_0$$

as follows from (1.35). Thus, the transverse electric field of a moving charge is stronger by a factor of γ in comparison with the same charge at rest. We have already come across it in Section 1.6. The component E_y causes a repulsive force

$$F_{y1} = qE_y = q\gamma E_0.$$

Let us now take into account the magnetic field. According to the formula (1.35) for B_z, we have:

$$B_z = \beta E_y = \beta \gamma E_0.$$

This field produces a Lorentz force which attracts the second charge to the first one. According to the expression (1.6) for the Lorentz force (See page 25), this force is equal to

$$F_{y2} = -q\beta B_z = -q\beta^2 \gamma E_0.$$

The net transverse force F_y can be found by adding the force of magnetic attraction F_{y2} to the force of electric repulsion F_{y1}:

$$F_y = F_{y1} + F_{y2} = q\gamma E_0 - q\beta^2 \gamma E_0 = \\ = q\gamma E_0(1-\beta^2) = \frac{qE_0}{\gamma}; \tag{2.1}$$

Just in case, note that

$$\gamma^2 = \frac{1}{1-\beta^2}.$$

2.1. CONTRACTING RODS

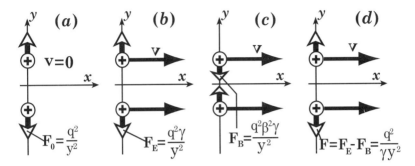

Fig.12. When two positive charges are at rest (a), the force F_0 of their electric repulsion obeys Coulomb's law. When the charges are set in motion (b), this electric repulsion becomes stronger by a factor of γ. But at the same time, there is a force of magnetic attraction (c) which eventually makes the net force of repulsion (d) weaker than in the case (a) of stationary charges.

All above-mentioned stages of the derivation of (2.1) are outlined in Fig.12.

For the charges at rest, the repulsive force is equal to qE_0. From (2.1) we see that, in spite of the magnetic attraction, the moving charges eventually do repulse each other. But the motion makes that repulsion γ times weaker. When the speed of the charges approaches that of light, the repulsion disappears altogether – the magnetic attraction and the electric repulsion counterbalance each other. After special relativity was developed, it became clear that law (2.1) applies not only to electromagnetic forces, but to all transverse interactions in nature. As for longitudinal forces, they remain unchanged provided the longitudinal distances between the interacting particles become γ times shorter. If those distances did not change, then the longitudinal forces would have become too weak to keep the parts of the body together unless the particles of the body had time to find and occupy their new equilibrium positions.

Transverse forces behave in a different way. All of them just become γ times weaker. What follows from that weakening? Let us focus on a particle which is in a state of equilibrium. The net force exerted on this particle on the part of all other particles is zero. Suppose that all forces of transverse interaction get reduced by a factor of γ simultaneously, irrespective of the positions of the particles.

Then the net transverse force must also reduce by a factor of γ. However, the net force was zero – as mentioned above. So, it will remain zero in spite of the motion. Therefore, the transverse positions of equilibrium of the particles, as well as the transverse dimensions of the body will remain unchanged.

But it is hard to believe that weakening of transverse forces does not affect in any way the behavior of particles inside of a solid. Suppose that a certain particle performing its thermal oscillations has moved off its position of equilibrium in a transverse direction. It will then be acted upon by a returning force, which will be γ times smaller in a moving body than in the same body at rest. Due to this relaxation, the thermal vibrations must slow down to some extent. So they will. And as the speed of motion approaches the speed of light, they will die out altogether. We'll return to this interesting question while studying the effect of uniform velocity on the tick of a moving clock in Section 2.3. Now we only note that the weakening of thermal vibrations does not affect in any way the transverse equilibrium positions about which the oscillations of particles occur and which determine the transverse dimensions of a solid body.

2.1.3. Revival of the Lorentz transformations for space

Measuring the length of a certain body is done by comparing it with the length of some ruler or tape-measure declared as a standard of length. But this standard, though manufactured in the best possible way, also shrinks by a factor of γ when it is set in motion through the ether. So, the result of the measurement depends on the choice of the standard used – whether it is in motion or at rest with respect to the ether. Since the moving standard is contracted by a factor of γ, the length of any body l', when measured by this standard, will be longer by a factor of γ than the length l of the same body measured by the standard at rest:

$$l' = \gamma l. \tag{2.2}$$

We may say that a body measured by an observer who is in motion is longer than the same body measured by an observer who is at rest. The word "observer" is somewhat misleading because nothing depends on the person who makes the measurement. This person may

2.1. CONTRACTING RODS

be in motion or at rest, sleeping or awake. Everything depends only on the standard – whether it is in motion or at rest. As for the observers, they can be replaced by automatic instruments which are able not only to measure, but also to record the results on a tape to be inspected by anyone later. Hence, speaking of an observers, we do not mean a person, but just the sets of instruments which are used for taking the measurements. As for the moving instruments themselves, they differ from the stationary ones, this difference being very great when the speed of motion approaches that of light. The speed of light c has entered (2.2) because the contraction of solid bodies conforms to the laws of electrodynamics where c is the most fundamental constant. As for the terms "moving" or "stationary", they are for the time being referred to the ether. We still "believe" that among all the moving rulers, there is one that is the longest and the most correct. It is the ruler which is fixed to the ether. All other rulers contract and lie.

Relation (2.2) can be extended by introducing two rectangular frames of reference: the one fixed to the ether, and the other one moving relative to the first along the x-axis with a constant speed v. Let x signify a distance between the reference point and the origin of the stationary frame (measured by a stationary ruler along the x-axis), and let x' be a similar distance between the same point and the origin of the frame in motion (measured by a moving ruler). Then

$$
\begin{aligned}
x' &= (x - vt)\gamma; \\
y' &= y; \\
z' &= z ,
\end{aligned}
\qquad (2.3)
$$

where $x - vt$ stands for the length l of the rod in motion, and x' stands for the length l' of the rod at rest. Time t is counted from the moment when the origins of the two frames coincide. The last two relations (2.3) show that transverse dimensions of moving bodies are not affected by motion.

Equations (2.3) are identical with the first three Lorentz transformations (see (1.11) and (1.12) on page 48). However, the primed quantities act here not as auxiliary mathematical variables that simply help us to solve Maxwell's equations, but rather as quite real physical quantities measured by moving instruments. Of course, they may also be used to simplify mathematical calculations, but this time we understand that x' is not a figment of our imagination. It represents the actual length measured by a moving ruler. We may say that we have succeeded in "reviving" the first three Lorentz transformations (1.11) and (1.12) by endowing their variables with a concrete physical meaning.

2.1.4. Length contraction of a moving rod in the eyes of an observer who accompanies the rod

Since all moving bodies undergo the Lorentz contraction, the observer moving together with them will not be able to notice this contraction. If the length of a certain body is measured by successive covering it with a meter stick, then the result of such measurement will not depend on whether the body and the stick are at rest or in motion. Some readers might wonder:

> "If I am on board a spaceship and this ship has accelerated to such an extent that the longitudinal dimensions have become, say, 4 times shorter, while the transverse dimensions have remained the same, then, with the contraction being so great, it might be tracked down without any measurements whatsoever. I could do it even with a naked eye; I would just look around the walls of the cabin and make a mental comparison between the length of the cabin and its width. Acting in the same way, namely with a naked eye, I would be able to track down the malfunction of the meter stick when seeing how it would contract while changing its orientation from transverse to longitudinal."

This argumentation is wrong. Measuring with the naked eye would not indeed introduce anything new as compared with using a meter stick or ruler for this measurement. When comparing the sizes with an eye, we in fact compare not the bodies themselves, but only their images on the retina. The retina in the eye, just like ourselves, contracts by a factor of γ as soon as it is set in motion. Therefore, no matter how hard we try, we can never detect any contraction of a spaceship's cabin with the naked eye, even if the spaceship, approaching the speed of light, would experience, say, a 100-fold contraction. You might say that this is an optical illusion, and you would be quite right. Because the moving ship really contracts, and this can always be registered by the stationary instruments, which are not moving with the cabin. The contraction can even be detected

2.1. CONTRACTING RODS

by the eye if this eye is at rest and its retina has not been affected by any deformation.[1]

The Lorentz contraction of the body remains invisible to the instruments moving together with the body not only after the body is accelerated, but even during the acceleration, provided the change of the velocity is gradual enough. If the acceleration is too great (but not so great as to destroy the body), the Lorentz contraction may occur with some delay. Everything depends on the swiftness of the particles of the body in taking their new equilibrium positions. If the velocity of the body changes slowly and gradually, the particles will have time to do it, and the length of the body at any moment of time will be determined not by the acceleration, but by the instant value of the velocity. If, on the contrary, the acceleration of the body is rapid enough, then the velocity may change before the particles have taken their new positions of equilibrium. In this situation, the Lorentz contraction will proceed not only during the acceleration, but also when the acceleration is over so as to give the particles a

---

[1] We should be very cautious when using an eye as an instrument for estimating the shape of a body moving fast relative to you. There may be some unexpected optical illusions which have nothing to do with relativity. If, for example, you watch a fantastic train passing by you with a speed near to that of light, your eye will currently take a snapshot of instant positions of the cars not at the moment of observation, but rather some time earlier because it takes a noticeable time for light to bring the images of the cars to your eye. The farther the car from you, the longer it takes to bring its image to your eye. If the train is approaching you, but has not passed by, its reflection in your eye will be greatly expanded. But if the train has already passed by you, its representation in the eye will be compressed. This effect has nothing to do with the Lorentz contraction and takes place when the measuring instruments (represented in our example by the naked eye) are confined within a small region of space. If we had a lot of instruments (such as rulers and clocks) distributed along the path of the train, they would show the true image of the train – it would experience only the Lorentz contraction and nothing else irrespectively of the fact whether it passed by our observation post or not. Amazing as they are, these optical illusions in the naked eye conceal the Lorentz contraction of the spherical moving bodies, such as stars, whose contours remain circumferential in spite of the Lorentz contraction, which turns a spherical body into a flat ellipsoid. It turns out that watching with the naked eye prevents one from seeing the Lorentz contraction rather than to simplify its detection. This optical illusions were discovered by J.Terrel – an American scientist – as late as 1959, so Einstein knew nothing about it.

certain time for finding and taking their new positions of equilibrium. This transitional period can be sensed not only by the stationary but even by moving instruments, because the moving instruments may give very odd and intricate readings during that period. This strange behavior will be caused by non-synchrony between their contraction and that of the accelerated body; even though, the final result for both the body and the instruments will be always the same – they will contract by a factor of γ. However, if the moving observer has overslept the acceleration and so did his instruments (they have been temporarily turned off), he will never be able to either measure the contraction, or even to establish the fact that it has taken place.

2.1.5. Michelson's experiment

Even though this entire story is in full agreement with Galileo's principle of relativity, it is difficult to concede the fact that the moving observer has no means to notice his own contraction. As the instruments at rest do register it, the contraction must be thought of as real, not imaginary. Perhaps the moving observer was not ingenious enough to register the contraction. Measuring with a rod as well as by eye has failed. But maybe it is still possible to invent another, more subtle, more refined or more sophisticated method that would be successful? Do we remember that distance can be measured not only with a rod or ruler, but also with a timepiece? It is this method that serves as a basis in radar facilities where the distance to an aircraft is measured by sending a radio signal and then waiting for its reflection to return from the aircraft. Having measured the time between the moments of sending the signal and its return, one can easily get the distance to the aircraft, because the speed of a signal is equal to the known value – the speed of light c.

Suppose we are on board a spaceship and want to use this radar method for detecting the Lorentz contraction of a massive metal plate placed in the cabin of the ship. To do it, we will install a source of light S on the plate as shown in Fig.13, and send two flashes of light from there simultaneously towards the mirrors M_1 and M_2. The first of them will propagate in the transverse direction with respect to the motion of the ship, while the second one will be directed along the motion. After reflection from the mirrors, the light signals return to source S and are registered there by a receiver that compares the duration of the round-trip of the two signals – to the mirrors and

2.1. CONTRACTING RODS

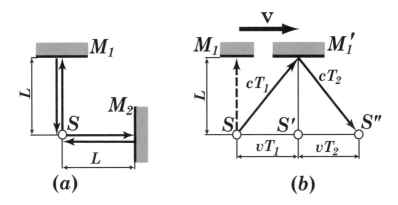

Fig.13. (a) The paths of two light rays in Michelson's interferometer which is at rest. The rays emitted by source S are reflected from the mirrors M_1 and M_2 and then return to the source to be compared with each other. (b) The path of the transverse ray of light in the moving Michelson interferometer. While the ray propagates from source S to mirror M_1 during time T_1, source S covers distance SS' and so does the mirror. While the ray returns from the mirror to the source during the time T_2, the mirror covers the distance $S'S''$. For clarity, the longitudinal ray is not shown here.

back. If the ship is at rest and the distances to the mirrors are the same: $SM_1 = SM_2 = L$, then the rays of light propagate as shown in Fig.13(a) and arrive at the receiver simultaneously. The time of the round-trip for each of the signals is equal to $2L/c$. When the ship moves with a speed v along the line SM_2, the Lorentz contraction takes place. The distance SM_2 shortens by a factor of γ and is equal to L/γ. It seems that for this reason the longitudinal ray of light should come back sooner than the transverse one. Let us check it up with taking into account the effect of the ship's motion on the trajectories of the light rays. Let us look first at the transverse ray. During the time T_1 which is needed by the ray to propagate from source S up to the mirror M_1 the ship with the mirror M_1 has time to cover the distance vT_1 along the motion as shown in Fig.13(b). The letters S and M_1 in the figure indicate the positions of the source and the mirror at the moment the ray of light is emitted, while the letters S' and M'_1 refer to the positions of the same objects at the moment when the ray arrives at the mirror.

When the ship is at rest, the length, covered by the ray of light to arrive at the mirror, is determined by the segment $SM_1 = L$. The ship being in motion, the light has to cover a longer path $SM_1' = cT_1$. Applying the Pythagorean theorem to the triangle, $SM_1'S'$, we arrive at the following equation:

$$(cT_1)^2 = (vT_1)^2 + L^2,$$

from which we obtain the desired time T_1 which is taken by the ray to reach the mirror M_1:

$$T_1 = \frac{L}{c\sqrt{1 - \frac{v^2}{c^2}}} = \frac{L\gamma}{c}. \qquad (2.4)$$

It will take the ray a certain time T_2 to get back. During that time the source S will cover the distance vT_2 and displace from point S' to point S''. The triangle $S'M_1'S''$ is equal to the triangle $SM_1'S'$ used above. Thus, the time T_2, spent on the way back, is equal to the time T_1, so that the net time T needed for the round-trip travel of the transverse ray is equal to

$$T = 2T_1 = \frac{2L\gamma}{c}. \qquad (2.5)$$

Thus, due to the motion of the ship, the time of the propagation of the transverse ray of light has increased by a factor of γ.

The same delay would happen to the sound signal, if the source S and the reflector M_1 were placed on a flatcar moving with a speed v, provided the speed of light c in equations (2.4) and (2.5) is replaced with the speed of sound in the air. If such an experiment was carried out within a closed railway car, there would be certainly no delay at all, because the air transmitting the sound would move together with the car. But in the case of a flatcar, the air does not take part in the motion and is felt by the moving instruments as a wind blowing against the motion. To return to the starting point, the sound wave has to propagate not only perpendicular to the wind but partially against it, which ultimately will result in the delay of the

2.1. CONTRACTING RODS

signal by a factor of γ. If the light is assumed to propagate through the stationary ether just like sound does it through the air, and the ether is assumed not to be dragged by the spaceship, then it will be the ether drift that may be declared to be the cause of the delay. Even though without any drift it is clear that the path $SM_1'S''$ in Fig.13(b) is longer than $2L$ (the hypotenuse being always longer than a leg), which is a sufficient reason for the light signal to increase the time of its travel under the action of motion.

The motion of our device will also affect the time of travel of the longitudinal ray. While this ray is propagating towards the mirror M_2 with the speed c, the mirror is running away from it with the speed v. Thus, the ray has to cover not only the distance L/γ that separates the mirror from the source at the moment when the ray is emitted (γ accounts for the Lorentz contraction of the plate), but also the way vt_1, covered by the mirror M_2 during the time t_1 of the ray propagation. Thus, the total way forth, covered by the ray, will be equal on the one hand to ct_1 and on the other hand to $(L/\gamma) + vt_1$. Having required of these two values to be equal to each other, we obtain the desired time t_1 taken by the ray to reach the mirror M_2:

$$t_1 = \frac{L}{\gamma} \frac{1}{c-v}. \tag{2.6}$$

With the growth of the speed v, this time increases because the mirror runs away from the ray faster and faster. This effect is partly compensated on the way back because this time the receiver of the signal moves against the ray, so that the distance covered by the ray shortens to the value $L/\gamma - vt_2$, where t_2 is the time of propagation of the ray backward. On the other hand, this very distance is equal to ct_2. Having required of these two values to be equal to each other, we get the desired time t_2, taken by the ray to return back:

$$t_2 = \frac{L}{\gamma} \frac{1}{c+v}. \tag{2.7}$$

Having summed up equations (2.6) and (2.7) with making use of (1.10) for γ (see page 47), we find the round-trip time t, taken by the longitudinal ray to travel forth and back:

$$t = t_1 + t_2 = \frac{L}{\gamma}\left[\frac{1}{c-v} + \frac{1}{c+v}\right] = \frac{2L}{\gamma c}\frac{1}{1-\frac{v^2}{c^2}} = \frac{2L\gamma}{c}. \quad (2.8)$$

Having compared it with the time T, determined by expression (2.5), we see the two times turn out to be absolutely the same in spite of both the ether drift and the Lorentz contraction. The times t and T, though proved to be γ times longer than in the case of a stationary spaceship, do not in the least differ from each other. The Lorentz contraction and the ether drift seem to have come to an arrangement to hide the fact of their motion from our instruments. Due to the Lorentz contraction the longitudinal ray spends on its propagation γ times less time than the transverse ray, but due to the ether drift it loses this gain in time. Eventually, the two effects cancel each other.

The comparison of the times of propagation of the two perpendicular rays of light on board a moving spaceship was made by Michelson and Morley in 1881. The role of a spaceship was played by our planet the Earth that is known to move about the Sun with the speed $v=30\,km/s$. The comparison of the two times was made by means of an optical device – interferometer, whose operation is based on the wave nature of light.

As we already know, the ray of light is in fact a traveling electromagnetic wave. Just like a sea wave, it has crests and troughs. The distance between two consequent crests is very small, it makes about 0.6 μm. This distance is called a wavelength. At the crests, the vector of the electric field is maximum, while pointing in a certain direction – for instance upward (if the wave is supposed to run in a horizontal direction.) At the troughs this vector is also maximum by magnitude, but its direction is opposite, for instance downward. A train of alternate crests and troughs moves with the speed of light c. If two rays come across each other at a certain place, for instance on a screen, the result of this meeting depends on how the crests and troughs of the two rays superimpose each other. If the crests of one ray coincide with the crests of the other, the merging of the rays will result in their intensification, and there will be a bright stripe on the screen. In the opposite case – with the crests of one ray coincident with the troughs of the other – the rays cancel each other and there will be a dark spot on the screen. If the two rays, emitted from the

2.1. CONTRACTING RODS

same source, take different paths and then meet again, the result of their merging will depend on how the ways covered by the rays relate to each other. If the ways are equal, then the crests of the two rays will coincide, and there will be a bright stripe on the screen. If the ways are 0.3 μm different, i.e. they differ half a wavelength, then instead of a bright stripe there will be a dark spot. If we are changing the length of the way of one of the two rays by displacing, say, one of the mirrors (see Fig.13(a)), there appear alternate bright and dark stripes on the screen. Multiplying the number of the changed stripes by 0.3 μm, we can measure with great precision the displacement of the mirror. Even though the displacement is less than 0.3 μm, it is still possible to measure it, because the bright and dark stripes interchange not instantly but gradually. Michelson's interferometer, being manufactured with perfect precision, allowed to detect the displacement of the mirror as small as 0.002 micrometers. With $v/c = 10^{-4}$ and the distance from the source to the mirror equal to 1.2 m, the Lorentz contraction made 0.006 micrometers, which exceeded about three times the smallest change in the length that could be registered by Michelson's interferometer.

Measurements were performed most thoroughly at different time of the day and in different seasons. At that time (in 1881) no one knew of the Lorentz contraction, and physicists searched for the ether drift, the existence of which was regarded as doubtless. In order to have this drift registered, the interferometer was gradually turned 90^0 and more, but no change whatsoever was registered in the interference pattern. Bright and dark stripes kept staying dead where they were, not displaying even a slightest tendency to interchange when the interferometer was being turned. The negative result of the experiment was explained much later, when Lorentz advanced his hypothesis on the contraction of the length of the moving bodies. After that, it became clear that the effect of the ether drift upon the propagation of the rays of light was fully counterbalanced by the Lorentz contraction and thus the negative result of the Michelson-Morley experiment was quite natural.

More than a century has passed since then. The measuring technique has improved. The Michelson experiment has been repeated many times, its precision always improving. But the result was always negative: no regular changes in the interference pattern have been registered.

So, what does Michelson's experiment prove? On the one hand it confirms the existence of the Lorentz contraction of the length of solid bodies by a factor of γ, this contraction being caused by the orbiting of the Earth round the Sun. If not for this contraction, the result of the experiment would have been positive. On the other hand, this experiment shows that the motion of the Earth does not affect the ultimate result of the measurements. In spite of the rotation of the device, the interference pattern does not experience even a slightest change – it behaves as if the device is at rest, and not in motion together with the Earth. The Lorentz contraction takes place, but it is impossible to have it registered with the moving instruments. If we did not know from astronomic observations that the Earth moves round the Sun, we would not have learned about it from the Michelson-Morley experiment, and our wonderful story of the Lorentz contraction and of the ether drift, compensating this contraction would be up in the air. We could always put it another way: "All those compensations are nothing but a figment of our imagination. In fact, the Earth is just at rest. The interference pattern stays dead not because the Lorentz contraction and the ether drift counterbalance each other, but because neither of them exists."

Not to get confused in this contradictory reasoning, it's time to sum up what we have known of the contracting rods. Ultimately, it may be reduced to the following objective regularities, which explain everything that was related above:

1. *A body moving with a constant speed v is contracted in the direction of motion by a factor of γ. That contraction can be registered and measured by the instruments at rest.*
2. *The uniform velocity does not affect the transverse dimensions of solid bodies.*
3. *The longitudinal forces inside a contracted moving body prove to be the same as in the stationary body.*
4. *The motion with a constant velocity weakens the transverse forces, acting inside the solid bodies, by a factor of γ.*
5. *The instruments which are in motion together with the body cannot register its longitudinal contraction.*

2.1.6. Beyond the light barrier

When the velocity of a body v approaches the velocity of light c, the value γ becomes enormously great. The body converts into a flat cake, perpendicular to the direction of its velocity. You might wonder: what would the body look like if its speed of motion exceeded c? In the next section we will learn that this question is absurd, because it is no way to accelerate any body to such a speed. But in this section we know nothing about it as yet. We are like little children who may be allowed not only to ask nursery questions, but also to look for reasonable answers.

Let us try to envision a body, for instance a rod, moving with a speed greater than that of light. What is its length equal to? If the speed v of the rod were equal to the speed of light c, it would then become absolutely flat and its length would be zero. But what would be if v exceeded c? Could it be that the length of the rod would become finite again? Let us try to calculate the value γ. The first surprise awaits us here. While calculating, we will have to extract a square root of a negative number. That is impossible. There cannot be a squared number that would be negative. A rod moving faster than light seems to have no length at all. What could it mean?

To answer this question let us peep again inside the rod and examine any of its particles. Until v was smaller than c, this particle was acted upon, in all directions, by forces of attraction and of repulsion that eventually canceled each other. Every time when our particle tried to run away, the balance of the forces was immediately violated and the particle was returned to its former position. That's why the rod could exist as a solid body and would not split into separate parts. All these forces were imparted through the electromagnetic field. But what would happen if the rod moved faster than the field? Then the particles that were behind would not be able to affect those which were at the front. Not a single particle would know anything of what was behind it. The particles that were at the front still would somehow affect those which were behind, while the particles that were behind would, on the contrary, fail to affect those which were ahead of them. This time the net force exerted upon any particle would not be zero any longer. It would always point backwards, trying to drive all particles to the speed-of-light barrier. The equilibrium would become impossible, and the rod would split into separate elements. How small would they be? Would we have spared anything of our rod?

Can it be that the rod will break into molecules or atoms? No, it will be not so easy for the rod to escape from the position in which it is trapped by the force of our imagination. No atom can be stable in the world lying beyond the speed-of-light barrier. As soon as the electron rotating round the moving nucleus happens to be in front of the nucleus, it will stop feeling it. Taking the opportunity, it will immediately leave its orbit, and the atom, just like the rod, will cease to exist. And what will happen next? Into what elements will the electron split? This time I am beginning to get angry. But take it easy. It is just a normal reaction of a person, who cannot answer a sensible question and has no desire to give way to his fantasy. Anyhow, the following can be said. Whatever is left of the rod, it will speed back with enormous retardation towards our world that lies before the speed-of-light barrier. It will return to the world of ordinary velocities, from which it could be extracted only in our imagination. The remnants of the rod will not be able to make a body whose length could be measured, or used to measure the length of any other body.

We have dwelled so long on this fantastic situation to show a special significance of the speed of light as compared with other speeds of motion that exist in nature. When the speed of a plane approaches the speed of sound, strong forces arise that hinder the plane in breaking through the sound barrier. But planes do overcome this barrier and develop supersonic velocities. At first, though, not everything went smooth. There were accidents when supersonic planes split into pieces while breaking through the sound barrier. But due to some improvements in the design and in the strength of the materials used, the supersonic velocities became normal in aviation. Planes normally do not disintegrate any longer when overcoming the sound barrier. That was achieved due to the fact that the particles of the hull and wings of the plane are held together not by vibration of the air through which the plain flies, but by the electromagnetic field that is able to propagate much faster than sound. If we assumed however that the plane moves faster than light, then the electromagnetic field would fall behind, so that there would be nothing to hold the particles of the plane together, and the plane would break. Even in vacuum the motion would become impossible if its speed exceeded that of light.

2.2. Inertia, energy and their unlimited growth

where we will see that an almost weightless Ping Pong ball, endowed with a good deal of electric charge, will acquire a considerable mass as soon as it is set in motion with a constant velocity. That mass will depend on the speed of ball's motion

2.2.1. What is the mass?

If a force is exerted upon a free body, the body experiences a change in its velocity. It leaves the state of rest (or the state of motion by inertia) and starts accelerating. Work is being done to be converted into a kinetic energy. For some reason, all material bodies do not like to have their velocities changed, and they do their best to oppose the process. That's why it is impossible to make an instantaneous change in the velocity or kinetic energy of any material body. Different bodies offer different resistance to the force that tries to change their velocity. Hence, under the action of the same force, different bodies acquire different accelerations. The ability of a body to offer resistance to the change in its velocity, or in other words, the inertia of a body, is characterized by a special physical quantity – the mass of the body.

The mass of bodies determines not only their inertia, but also the forces of mutual attraction, experienced by them according to the law of universal gravitation. By means of precise measurements, it was established that the inertia of bodies and their mutual attraction are in full correspondence with each other. The greater inertia of the body, the greater the force of its gravitational interaction with some other body, e.g. with the Earth. That's why under the action of the force of gravity all bodies get the same acceleration irrespective of their mass. The larger the mass of a body, i.e. the greater its inertia, the stronger is its attraction to the Earth, so the acceleration is always the same. This is generally put like this: *The inertial mass of any physical body is equivalent to its gravitational mass.*

Though such wording elucidates the role of mass, yet it does not give a definite answer to the question what, exactly, the mass is. To give a definition to a physical quantity means to specify how that quantity could be unambiguously measured. This is the only possible way in physics to set up a reasonable definition. So, how are we to define (to measure) the mass of a body? We can of course agree that the mass is a constant of proportionality between the force and the acceleration in Newton's second law, or, which is the same, that the mass of a body is defined as the acceleration, acquired by the body under the action of a unit force. This definition sounds well. But what will be the answer to the question what the force is. A force is defined as the acceleration acquired by the body of a unit mass. It is felt that something is wrong here. When we define the mass, we use the force as a given concept. And when defining the force, we use the mass as a starting point of our definition. We are entrapped into a vicious circle. To get out of it, we have either to define the force without using the concept of mass, or to define the mass without using the concept of force. Fortunately, nature allows to do this. The second way seems more convenient. But how can we exclude force from the definition of mass? This can be done by using the two laws simultaneously: Newton's second law

$$F = ma,$$

and the law of universal gravitation for two identical masses interacting with each other

$$F = \frac{Gm^2}{r^2};$$

where F is the force, m – the mass, a – the acceleration, r – the distance between the interacting masses, and G is the gravitational constant. What if we exclude force F from these two laws, so that ma and Gm^2/r^2 will be left in privacy together? We will then get:

$$ma = \frac{Gm^2}{r^2} \quad \text{or} \quad m = \frac{ar^2}{G}$$

and there we are. If formulated verbally, it sounds like this: *The mass of a given body is defined as the acceleration, imparted by this body to another body identical to it, placed a unit distance apart.*

We have escaped from the vicious circle – no forces take part in this definition. The very fact of the possibility of such a forceless definition is suggestive. Nature itself whispers a prompt into our ears: "Since acceleration is due to nothing but gravity, we can do even without any forces." Later on, in Section 2.9, we will make use of this prompt. So, now let us focus on inertia and put aside, for the time being, the gravitational forces as well as an alluring possibility of expelling them from the list of physical notions.

2.2.2. How the mass is created

Now, that we have managed, though with some efforts, to answer the question: *"What is the mass?"*, there arises another, more challenging question: *"What is its origin?"* We should concede at the outset that the modern physics is unable as yet to give an unequivocal answer to this question. But in some particular cases, the mechanism of mass origination is fairly evident, and we will not hesitate to make use of it.

Imagine a Ping Pong ball of mass M which is given a good amount of electric charge q of any sign. If this ball is placed into an external uniform electric field E_0, it must accelerate in the direction of that field. It seems that the acceleration of the ball should be equal to the force $F = qE_0$ divided by the mass M. So it is, provided the charge q is small enough. But if this condition is not met and the amount of charge is too great, then the inertia of the ball may prove many times higher due to the electromagnetic processes that follow the ball's acceleration. The clue to this additional inertia is furnished by the growth of the magnetic field in the process of the ball's acceleration – the faster the charge moves, the greater its magnetic field, whose lines form a familiar bunch of rings strung onto the trajectory of the charge. In Fig.14 those lines lie in the planes perpendicular to the plane of the drawing and are seen as small circles with either a dot or a cross inside. The dot reminds the tip of an arrow and thus indicates the point where the magnetic line comes out of the drawing, while the cross is like the tailpiece of an arrow and indicates that the magnetic line points into the figure. These magnetic lines form a magnetic flux which encircles the trajectory OO' of the ball.

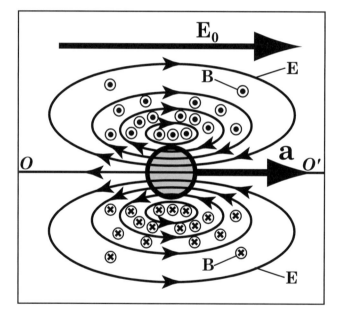

Fig.14. A positively charged Ping Pong ball is accelerating along a straight line OO' from left to right. The moving ball is a current whose magnetic lines encircle its trajectory OO' in the planes perpendicular to the plane of the drawing. They are represented by small circles which contain either a dot, showing that the magnetic field points out of the plane of the figure, or a cross, showing that the magnetic field points into the plane of the drawing. Due to the increase in the speed of motion, the magnetic flux is growing and generating the vortical electric field E, whose lines of force are represented by closed solid lines. This field offers resistance to the ball acceleration.

When the ball is accelerating, this flux is growing and, in accordance with the law of electromagnetic induction, there appears a vortical electric field **E**, whose lines of force are outlined in Fig.14. Obeying Lenz's law, that field is directed against its cause, i.e. against the acceleration **a** of the ball. Because the electric field is vortical, its lines of force cannot end at the ball – they must pass through the ball and form closed loops. In addition to this vortical electric field, there is also an ordinary associated electric field of a charge moving with a constant velocity. This field undergoes the Lorentz contraction and was shown in Fig.11. (See page 56.) Its lines of force originate on the ball and extend symmetrically both forward and backward, so that the resultant force, acting on the ball on the part of this field, is zero. This field does not affect the acceleration of the ball and, for the sake of clarity, is removed from Fig.14.

Thus, the acceleration of the ball takes place under the action of two electric fields – the external field E_0, which pushes the ball ahead, and the ball's own vortical counter-field E, created according to the induction law and pushing the ball backward – against the external field. The acceleration a of the ball can be found from Newton's second law:

$$q(E_0 - E) = Ma; \quad \text{or} \quad qE_0 = Ma + qE. \qquad (2.9)$$

There are two forces in this equation: the external force $F = qE_0$, provided by external sources, and the force qE created by the ball itself by means of its own field. Since the second force is produced by the ball and is directed against the acceleration, we have moved it to the right part of the equation (2.9), placing it side by side with the familiar item Ma. If the ball moved with a constant velocity and its magnetic field was just propagating in space without changing its magnitude, then the ball's own electric counter-field shown in Fig.14 would not arise at all. But once the ball starts accelerating and its magnetic field begins to grow, there appears a retarding vortical field E which is proportional to the amount q of the charge and its acceleration a. We can write that $E = kqa$ where k is a constant of variation that depends, in particular, on the size of the ball. Substituting this expression for E into (2.9), we arrive at

$$F = qE_0 = (M + kq^2)a. \qquad (2.10)$$

Two masses are involved in this equation: the ordinary mass M and the electromagnetic mass $m = kq^2$. If the ball is light enough or the charge q is large enough, then the electromagnetic mass $m = kq^2$ can prove much greater than the ordinary mass M. A ball of ordinary mass M (say, a few grams) can acquire an electromagnetic mass as large as, say, ten kilograms. Of course, there is no question of playing Ping Pong with such a ball. In this situation, electromagnetic induction is responsible for almost all the mass of the ball in accordance with the following approximation:

$$F \cong kq^2 a = ma.$$

2.2.3. Mass-velocity dependence

Does the speed of motion affect the electromagnetic mass $m = kq^2$? It certainly does not affect the charge q. In physics, the charge is conserved always and everywhere. There remains factor k. It depends on the size of the ball – the smaller the ball, the larger its electromagnetic mass. This is explained by the pattern of the vortical electric field whose lines of force converge at the ball before passing through it as shown in Fig.14. Thus, the smaller the ball, the stronger its own counter-field at the place of its location, and the larger its electromagnetic mass. The formula for the electromagnetic mass of a charged sphere of radius R was derived as far back as at the times of Lorentz. In the CGS units it has the following form:

$$m_0 = \frac{q^2}{6\pi c^2 R}. \qquad (2.11)$$

The ball's radius R is in the denominator, which confirms the growth of the electromagnetic mass of the ball with reduction of its size.

The previous section tells us that, with the growth of the speed, the ball undergoes the Lorentz contraction and becomes shorter by a factor of γ. Because its transverse dimensions do not change, it turns into an ellipsoid. The ball becomes smaller, though only in the direction of motion. Equation (2.11) suggests some growth of the mass without giving the exact value. According to the strict derivation (also made at the times of Lorentz), the mass is growing exactly as γ.

2.2. INERTIA, ENERGY AND THEIR UNLIMITED GROWTH

$$m = m_0 \gamma = \frac{m_0}{\sqrt{1 - \frac{v^2}{c^2}}}. \qquad (2.12)$$

Lorentz firmly believed in this formula in spite of its being in conflict with the experimental data available at the time. The experiments were made with the fast electrons, arising through a radioactive decay. Later on, it turned out that the measurements had not been exact.

The mass-velocity dependence of the ball is explained by the Lorentz contraction of the electromagnetic field, that accompanies the ball and is shown in Fig.11. (See page 56.) If even the ball itself did not suffer any contraction caused by its motion, the field of the ball would all the same have contracted and the mass would have increased, though not exactly by a factor of γ. Why does the contraction of the field increase the mass? The clue is given by the growth of the energy of the field in the process of the field contraction. When the field is shrinking, the density of its energy increases as γ^2 or so, while the volume occupied by the field decreases as $1/\gamma$ or so. Eventually the net energy of the field grows exactly by a factor of γ. It turns out that the closer to the speed of light, the higher cost is paid for any further increase in the speed of a body. For example, a 10-fold raise in the energy may be needed in order to increase the speed of a body just by 1%. This brings us to the situation when even a tiniest increase in speed can be achieved only through a lot of force to be applied. In other words, the mass of the accelerating body proves extraordinarily large.

With the speed v approaching the speed of light c, the mass of a body, according to (2.12), tends to infinity. What does it mean in practice? Humanity has overcome many barriers while paving the way to the technological progress – the sound barrier being a good example of it. However high these barriers were, they usually proved passable. The light barrier seems infinitely high because we do not see any real way of overcoming it. The closer we approach this barrier, the higher it becomes. Therefore the speed of light is now assumed as the ultimate limit on speed of any moving body carrying some energy or information.[1]

You can disagree:

"Is it reasonable to arrive at such a pessimistic conclusion on account of the properties of a miserable toy-like ball, leaving alone that it is artificially electrified? Discharge the ball and it will become as light as ever. When being accelerated, it will, though, experience the Lorentz contraction, but being electrically neutral, it will not be involved in the mass-velocity dependence."

This reasoning is wrong. Even when the ball is neutral, it contains a lot of various electromagnetic fields inside. After the acceleration, not only the ball itself will be contracted, but all those fields as well. The energy of the internal fields will all the same increase by a factor of γ, and, according to the law of electromagnetic induction, all the same there will be electric counter-fields inside the ball, which will vindicate (2.12)

The relation (2.12) is reliably proved by experiments and widely used in engineering. It would be impossible to design a high-energy charged particle accelerator without taking into consideration the mass-speed dependence.

2.2.4. Mass-energy equivalence

As soon as Einstein universalized the relation (2.12), wonderful corollaries of global importance emerged from it. Let us for instance answer the following question: what mass will be imparted to a certain body when this body, initially at rest, is set in motion at a certain uniform speed v – very small in comparison with the speed of

---

[1] There is, however, at least one exception to this rule. When the force exerted on a charged particle, say on the electron, is so strong that the particle has no time to rearrange its internal structure ("to undergo the Lorentz contraction") and acquire the current mass corresponding to the increasing current value of its speed, then the particle has time at least to reach the speed of light. Something like this happens when the electron and the positron (a copy of the electron but positively charged) turn out to be in contact with each other. Both of them convert into light. They manage "to climb the barrier" but fail to overcome it. When these two particles "touch" each other, the instantaneous force of their interaction proves so large that they have time to reach the speed of light sooner than the path covered by them exceeds their diameter. The electromagnetic force greater than in this example is unknown in nature.

2.2. INERTIA, ENERGY AND THEIR UNLIMITED GROWTH

light? (2.12) tells us *how many times* the mass is increased – it is increased γ times. In order to know *by what value* the mass is increased, we have to transform (2.12) by borrowing from mathematics the following approximate rules:

$$\sqrt{1-x} \cong 1 - \frac{x}{2} \qquad \frac{1}{1-x} \cong 1 + x \ . \qquad (2.13)$$

These approximations are good enough when x is much smaller than unity. You can make sure of it for yourself by substituting test numbers for x. Making use of the approximations (2.13), we can transform (2.12) in the following way:

$$m = m_0 \gamma = \frac{m_0}{\sqrt{1-\beta^2}} \cong \frac{m_0}{1-\frac{\beta^2}{2}} \cong m_0 \left[1 + \frac{\beta^2}{2}\right] =$$

$$= m_0 + \frac{m_0 v^2}{2c^2} \ . \qquad (2.14)$$

This formula is good in clarifying the two parts of the mass one of which (m_0) is a *rest* mass and has nothing to do with motion, while the other one $(m_0 v^2)/(2c^2)$ is the mass increase produced exclusively by motion. Look at this increase attentively. It is only factor c^2 that makes it different from the formula for the kinetic energy $(m_0 v^2)/2$, whose derivation in mechanics is familiar to us. Though the factor c^2 is very large, if the time is measured in seconds and the length – in meters, it does not matter, because c turns into unity when distance is measured in light seconds.[1] Then the speed of light $c = 1$ and is dimensionless. If however $c = 1$, then the mass increment of a body is just equal to the increment of its energy. Thus, the increment of mass and the increment of energy are identical physical variables, that become apparently different if only length and time are measured in different units. But if the increments of two quantities have proved identical, doesn't it mean that the quantities themselves are identical too? How should we deal, however, with the

---
[1] A light second is a distance covered by light in one second. Astronomers use such units all the time, though they prefer light years.

addend m_0 in (2.14)? It looks like the mass of a body which is at rest. What energy does it correspond to?

Einstein declared that any body at rest contains the energy W_0 equal to its rest mass m_0 multiplied by c^2.

$$W_0 = m_0 c^2. \qquad (2.15)$$

On certain conditions, this energy can be released and can perform work. Let us take for example an automobile engine. When a reaction of combustion takes place in the engine and some energy is released, which makes the car move, a certain, albeit very small, amount of the rest mass of molecules of petrol is spent. If all the secondary combustion products are thoroughly weighed instead of being sent to the exhaust, then their total rest mass will prove a little bit less than the total rest mass of the molecules of petrol, burned down in the engine. And that "little bit", multiplied by c^2, is quite enough to make the car move The law (2.15) served as a starting point for utilizing the nuclear energy. Experiments with microscopic amounts resulted in splitting the nucleus of uranium into two lighter nuclei. The net mass of these nuclei proved a bit smaller than the mass of the initial nucleus. That gave rise to the idea of releasing and utilizing the nuclear energy. And that "bit" multiplied by c^2 was sufficient to make the bomb.

The law (2.15) refers to the body at rest and is a particular case of Einstein's general equation relating to any material body which is not necessarily at rest:

$$W = mc^2. \qquad (2.16)$$

Here m is the mass of the body exceeding its rest mass by a factor of γ, whereas W is the net energy of the body which comprises the rest energy $W_0 = m_0 c^2$ and the kinetic energy. The physical meaning of the value γ, that we have frequently come across, is now becoming clearer. It represents the net energy of a body in motion related to the energy of the same body at rest. If we assumed the rest energy as a unit of the net energy in motion, then γ would be just the net energy of the body moving with a speed v.

Equation (2.16) is called the principle of energy-to-mass equivalence. But what is meant by that equivalence? It is sometimes inappropriately interpreted as a possibility for the energy to be converted into mass and vice versa. What is true is that mass and energy are just two identical physical quantities that may differ from each other in the way of measuring and in the choice of units. If they are measured in the same units, then $c = 1$, and the equation (2.16) turns into an identity: $W = m$. But if mass and energy are equal to each other, why do they have different names and designated with different letters? There is only one reason for doing so – they are often measured in different ways and represented in very different units. The value $W = mc^2$, if measured in the CGS units, is large enough to make that difference quite impressive. But whenever it is convenient, physicists measure the mass of elementary particles in units of energy, and there is never any misunderstanding about it.

2.2.5. *Relativistic dynamics*

Let us return to the formula (2.12) in order to see how the mass-velocity dependence affects the dynamics of bodies – how it modifies the connection between the force and the acceleration acquired under the action of that force. First of all, let us surface a surprising thing. Though mass-velocity dependence was quite new and unexpected property from the standpoint based on classical mechanics, Newton's second law did not suffer from it even in the least. Even nowadays, it sounds exactly in the same way as it did at the time of Newton: *When a free body is acted upon by a constant force, the rate of change of its momentum is equal to the force*:

$$\mathbf{F} = \frac{d(m\mathbf{v})}{dt}. \tag{2.17}$$

Note that velocity $\mathbf{v}$ and force $\mathbf{F}$ are vectors. It is hard to say what namely made great Newton formulate his law in that form. He could have easily written

$$\mathbf{F} = m\frac{d\mathbf{v}}{dt}$$

placing mass first – before the derivative. That would have seemed simpler. As for the mass-speed dependence, no one had ever suspected of its existence either at that time or even two centuries later. Nevertheless Newton had chosen the form that needn't to be modified formally even now, after revolutionary changes in physics were made by special relativity.

Let us see now what follows from Newton's second law if we take into account that mass depends on the speed of motion. Of course, Newton himself did not know anything about these corollaries. First of all, let us split the right part of (2.17) into the two vector items:

$$\mathbf{F} = m\frac{d\mathbf{v}}{dt} + \mathbf{v}\frac{dm}{dt} = m\mathbf{a} + \mathbf{v}\frac{dm}{dt} \ . \qquad (2.18)$$

The first addend here is familiar to us. As for the second addend, it is an outgrowth of the mass-velocity dependence. In other words, during the time dt, the momentum $m\mathbf{v}$ is changing for two reasons — first, due to the change of velocity $\mathbf{v}$, which is taken into account through the first addend, and secondly, due to the change of mass m (caused by the change of the speed $\mathbf{v}$), which is represented by the second addend.

Let us see how the second addend affects the acceleration of the body. We will begin from a special case with the force $\mathbf{F}$, applied to a body perpendicularly to its velocity $\mathbf{v}$. Suppose we want to impart a transverse acceleration to a body which is already moving in the longitudinal direction. However, the time interval dt is so small, that the velocity $\mathbf{v}$ has time to change a bit only its direction, but not the magnitude. Because the magnitude of the velocity remains unchanged, so does the mass in accordance with (2.12), where there are no vector quantities. Thus, $dm = 0$, and Newton's law (2.18) reverts to its ordinary form:

$$F_\perp = ma_\perp . \qquad (2.19)$$

Symbol "$\perp$" reminds that the force and acceleration are *perpendicular* to the velocity $\mathbf{v}$. We have arrived at *the first rule of relativistic dynamics*:

> *If a moving body is acted upon by a force perpendicular to its velocity, the body is accelerated in the direction of the force, and the constant of proportionality between the force and the acceleration is equal to the mass of the body.*

2.2. INERTIA, ENERGY AND THEIR UNLIMITED GROWTH

Formally, this rule does not bring about anything new, but we must remember that the constant of proportionality $m = m_0\gamma$ depends on the speed of the body.

Let us switch over to the second case. Let the force **F** act along the line determined by the vector **v**. In other words, the body is either accelerated or retarded in the direction of its motion. This time, the velocity would change in magnitude, and the second item in the right part of equation (2.18) must be different from zero. To find that item, we have to resort not only to algebra, but also to the rules tabulated in calculus handbooks. To begin with, let us shape (2.18) into a form which lends itself to further evolution:

$$F = ma + v\frac{dm}{dt} = ma + v\frac{dm}{dv}\frac{dv}{dt} = ma + va\frac{dm}{dv}, \quad (2.20)$$

where for the sake of simplicity all the vectors were turned into scalars — this is permissible because in this particular case all vectors change only in their magnitude. To pave the way for further derivation, we have to use the mass-velocity dependence (2.12) in order to find the derivative dm/dv:

$$\frac{dm}{dv} = m_0\frac{d\gamma}{dv} = m_0\frac{d}{dv}\frac{1}{\sqrt{1-\frac{v^2}{c^2}}} = \frac{m_0 v}{c^2\left[1-\frac{v^2}{c^2}\right]^{\frac{3}{2}}} = \frac{m_0\gamma^3 v}{c^2}. \quad (2.21)$$

To finish our derivation, we have to make two substitutions in (2.20) — placing the right-hand parts of (2.21) and (2.12) instead of dm/dv and m respectively:

$$\mathbf{F} = m_0 \gamma a + \frac{m_0 v^2 \gamma^3 a}{c^2} = m_0 \gamma a (1 + \beta^2 \gamma^2) =$$

$$= m_0 \gamma a \left[1 + \frac{\beta^2}{1 - \beta^2} \right] = \frac{m_0 \gamma a}{1 - \beta^2} = m_0 \gamma^3 a \ . \tag{2.22}$$

There were many derivations, but the result proved very simple. It is *the second rule of relativistic dynamics:*

If a moving body is acted upon by a force applied in the direction of motion, the body is accelerated in the direction of the force, and the constant of proportionality between the force and the acceleration is equal to $m_0 \gamma^3$.

Though the variable $m_0 \gamma^3$ is not strictly a mass (because, for instance, it is not equivalent to the net energy of the body), it is often called a longitudinal mass to distinguish it from the true mass $m_0 \gamma$, which is sometimes called a transverse mass. Because the longitudinal mass is proportional to the cube of the net energy γ^3, it grows very fast when the speed of the body approaches the speed of light. The light barrier turns out to be even more unreachable than it seemed above.

Now we are passing on to the third, most interesting rule of relativistic dynamics. Suppose a force is applied to a moving body at an arbitrary oblique angle with its velocity. This angle is neither zero, nor 90°. Let us rewrite equation (2.18), having rearranged it into an explicit formula for acceleration **a**:

$$\mathbf{a} = \frac{\mathbf{F}}{m} - \frac{\mathbf{v}}{m} \frac{dm}{dt} . \tag{2.23}$$

Explicit as it is, this expression is of no avail to us in calculating value **a**, because dm/dt still remains unknown. However, we can be sure that this derivative is different from zero. It cannot be other

2.2. INERTIA, ENERGY AND THEIR UNLIMITED GROWTH

way because, this time, the speed of the body does change, and so does the mass. Let us try to draw some conclusions about the direction of the acceleration **a** without cumbersome calculation of dm/dt.

Equation (2.23) includes three vector variables. One of them is subtracted from the other to give the third vector – the acceleration **a**. The vectors that take part in the subtraction make some angle with each other (force **F** makes a certain angle with velocity **v**), and both of them are different from zero. Hence the three vectors in equation (2.23) form a vector triangle shown in Fig.15. This looks like

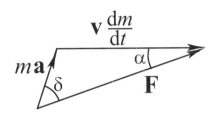

Fig.15 When force **F** is applied at an oblique angle α to a body, moving with a velocity **v**, the acceleration **a** is directed surprisingly not along the force, but makes a certain angle δ with it.

resolving a certain force into a pair of its components and brings us to *the third rule of relativistic dynamics*:

> ***If a force is applied to a moving body at an oblique angle to the velocity of the body, then the direction of the acceleration does not coincide with the direction of the applied force.***

It turns out that all the material bodies, even in absolute emptiness, behave like sailboats, whose acceleration is not necessarily in the direction of the wind. Had Lorentz noticed this effect, he would have regarded it as an excellent proof of the existence of the ether. But in fact it is not helpful in detecting the ether. On the contrary, it helps the ether to cover up its traces and to cancel other phenomena that otherwise might display the ether drift. An example of such a situation will be given in the last paragraphs of Section 2.8.

When using the third rule, we should remember that it refers to an acceleration, and not to a change in the momentum, which is always directed strictly along the force in full accordance with Newton's second law. This proves once again that all the three rules of relativistic dynamics, though they look like innovations, have been obtained not in spite of Newton's laws, but due to them.

We now have enough material to approach the most sacramental question in special relativity – the effect of motion on the flow of time.

2.3. Stopping time

where we will see that every clock, whatever its design might be, has its own good reasons to slow down its ticking after it is set in motion with a constant velocity

2.3.1. Where is time stored?

To make sure of the reality of space, it is enough to measure it — by taking steps or with a standard rod, with a tape-measure or with a range-finder, with sliding calipers or with a micrometer, with a radar or with an interferometer. If we were not able to accomplish those measurements, there would be no room for space in a physical theory. The same refers to time. Time may exist in physics as long as it can be measured. It even seems that if there were no clocks, then there would be no time.[1]

Clocks should be understood here in the utmost broad sense. Time may be counted by means of any periodical, repeating event: sunrise or sunset, the beating of the pulse, the revolution of clock-hands, the oscillations of a pendulum, the rotation of an electric motor, the rings on a cut down tree, etc. All these phenomena can be used as clocks, each having its own rate and precision. To answer the question whether the motion with a constant velocity affects time, it is necessary to investigate the effect of the uniform velocity upon the tick of various clocks.

---

1 At the second thought, this aphorism needs an essential refinement. In fact, nature hides the World Time somewhere very deep while the clocks reproduce it only approximately — as precisely as they can. See some tips in the last paragraphs of the historical review — at the very end of the book, pages 269-270.

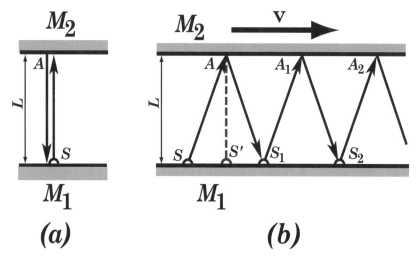

Fig.16 An outline of a light-ray clock, consisting of two parallel mirrors M_1 and M_2 of indefinite width, with a single short flash of light reverberating between the mirrors. The flash of light is emitted by the source S attached to mirror M_1. In the left drawing (a) the mirrors are at rest in contrast to the right drawing (b) where both mirrors are moving to the right with a speed v together with the source. It is seen how motion increases the path traveled by the flash.

2.3.2. Light-ray clock

A good stock of knowledge that we acquired in the preceding sections is sufficient to consider a rather broad assortment of various clocks. *A light-ray clock* is the simplest of them. Imagine two parallel plane mirrors M_1 and M_2 shown in Fig.16(a) and a source of light S that, at a certain moment of time, has emitted a single short flash of light from mirror M_1 towards mirror M_2. The flash covers the distance L between the two mirrors in the time L/c and arrives at the point A on mirror M_2, after which it is reflected back to mirror M_1 and returns to S. The round-trip time is equal to $2L/c$ as counted from the launch of the flash. If the mirrors are at rest, the flash appears at point S periodically every $2L/c$ period of time. If a counter is installed at point S, its readings will indicate the time passed since the launch of the flash. The precision of measuring the time by this clock is determined by the time interval $2L/c$ between two successive ticks of the counter.

2.3. STOPPING TIME

Let us now see how this light-ray clock will change the frequency of its tick after it is set in motion with a uniform velocity v. If the direction of motion *is coincident with the planes of the mirrors*, then the flash of light propagates along the broken line $SAS_1A_1S_2A_2$ as shown in Fig.16(b). Let T be the round-trip time spent by the flash to travel to mirror M_2 and to return back. During this time, the counter S moves off the distance $SS_1 = vT$. For this reason, the path $SAS_1 = cT$ covered by light proves to be larger than $2L$, because in the right triangle SAS' hypothenuse SA is longer than the leg AS'. By means of Pythagorean theorem it can be shown that the motion makes the path of light γ-fold longer. (A similar derivation was made in Section 2.1 in connection with the analysis of the Michelson experiment.) The time interval between two successive ticks of the counter will also increase at the same ratio. Hence the motion with a uniform velocity makes the tick of a light-ray clock slower by a factor of γ.

Let us change the direction of clock's motion. Let it move not parallel, but perpendicular to the mirrors. Though the trajectory of the flash of light is quite different, the tick of the clock will be again slower by a factor of γ. While light is propagating from M_1 to M_2, mirror M_2 is running away, which will make the path of the flash longer. On the way back, when the flash of light is returning to its source, mirror M_1 is moving against the light, which will shorten the path traveled by the flash. But the lengthening of the path on the way forth proves to be more significant than its shortening on the way back, and eventually the net round-trip path will again prove to be γ-fold longer in comparison with the case of a stationary clock. This derivation was also carried out in connection with the Michelson experiment. (See equations (2.6)-(2.8) on pages 89,90.)

2.3.3. Is a light-ray clock an exception to the rule?

Thus, the light-ray clock in motion ticks γ times more slowly than the same clock at rest, irrespective of the direction of motion. The slowdown in the tick depends on how close the speed v is to the speed of light c. If v approaches c very near, then the clock almost stops ticking. It might be suspected that such a strange behavior is characteristic of only a light-ray clock, while "ordinary" clocks can stand the motion without any consequences. The light ray as an inevitable element of a light-ray clock is a good candidate to blame for all the tricks that are performed by this clock in the vicinity of the light barrier. A sound-wave clock, for example, is also very tricky in the vicinity of the sound barrier with the speed of sound in the air 320 m/s standing for the speed of light. Indeed, let us imagine that the source S in Fig.16 emits a short burst of sound which propagates in the air, reverberating between the two plane sound reflectors M_1 and M_2. If such a sound-wave clock is moving with a constant velocity through the stationary air, the sound, before returning to S, will have to travel a longer path, as shown in Fig.16(b), and the ticking of the clock will be γ times slower, provided the speed of sound is substituted for the speed of light in the expression (1.10) for γ (See page 47). If such clock is moved with the speed of sound, it will stop ticking altogether. The source S will be running away from the sound signal so fast that the signal will never be able to return back to S. And what if the speed of motion is supersonic? How will the clock behave in that case? The sound wave will be blown away by the air wind, and the counter will never register anything. Such a clock should be given up as either showing the wrong time when moving more slowly than the sound, or failing altogether when its speed becomes supersonic. Can it be that the light-ray clock should be given up too?

In fact, there is a great difference between a sound clock and its optical analog. The slowdown of the tick of a sound clock, which is caused by its motion through the stationary air, is removable. To make the sound clock invulnerable to relativistic tricks, it is sufficient to place it inside an air-tight hood. The air will then move together with the reflectors, so that the air wind as well as the slowdown of the tick will become impossible. The clock will retain its former rate and will show the true time even when the speed of motion is supersonic. Not so for a light-ray clock. Whatever hood might be used to cover the mirrors, that would not protect the rays of light from the influence of motion. There is no such an enclosure that could protect the clock from an "ether wind", i.e. from the lengthening of the path of the light flash shown in Fig.16. And even the sound-wave clock protected by a hood, when accelerated up close to the speed of light, will get its tick slower by a factor of γ – exactly as much as a light-ray clock. At first thought it might seem strange. Isn't this clock a sound-wave? If yes, then what does it have common with light? How can light affect the propagation of a sound wave? It turns out that light not only can but even must do it quite successfully. Let us recall that a sound wave consists of condensations and rarefactions of the air, propagating with a speed of 320 m/s. To make these condensations propagate, molecules of the air must somehow interact with each other. This interaction is realized through the electromagnetic fields of the molecules flying up close enough to each other. But it is not the same to the electromagnetic field whether the interacting molecules are fixed to the ether or all of them are moving through it with a speed approaching the speed of light. So, the electromagnetic field also participates in the formation and propagation of a sound wave. Because light propagates about a million times faster than sound, the electromagnetic change of the velocity of sound in a sound clock is usually negligible. But if the speed of this clock approaches the speed of light, the effect of electromagnetic processes on the propagation of the acoustic wave becomes not only noticeable, but even dominant.

2.3.4. A clock with a spring-driven pendulum

The electromagnetic field and its inalienable characteristic – the speed of light – make their investment into the work of clocks of any design, including even those that seem to have nothing to do with electrodynamics. Let us take, for example, an ordinary alarm clock of a most ancient, mechanical design. Its main part is a spring-driven pendulum, whose frequency of oscillations determines the clock's rate. The spring-driven pendulum, if simplified, can be envisioned as a load, suspended by an elastic spring. If we pull this load down and then let it go, it will oscillate vertically, the spring alternately contracting and extending. The frequency of the oscillation depends on the mass of the load m and on the modulus of elasticity of the spring k. The latter is defined as a constant of proportionality between the force exerted on the spring and the length of its extending or contracting. Let us now muse about the origination of elasticity. Where does it come from?

When the spring is extending, its particles are shifting from their equilibrium positions, where the net forces acting on them were balanced. All these shifted particles are acted upon by electric and magnetic forces that try to return them to their equilibrium positions. This makes the extended spring contract when you let it go. Though, in our example, the spring does not experience the Lorentz contraction (it is oriented perpendicularly to its motion) and the equilibrium positions of its particles are unaffected by motion, the elastic forces (arising in the process of the spring's oscillations) do depend on the motion and are γ times weaker. This makes elasticity inversely proportional to γ: $k \not\propto 1/\gamma$. Besides, the mass of the ball increases by a factor of γ according to the first rule of relativistic dynamics : $m \not\propto \gamma$. The frequency of oscillations f of the load, suspended by a spring, is known to be proportional to the square root of the elastic modulus of the spring related to the mass of the load:

$$f \not\propto \sqrt{\frac{\kappa}{m}} \not\propto \frac{1}{\sqrt{\gamma^2}} = \frac{1}{\gamma}.$$

Because the ratio k/m proved inversely proportional to γ^2, the motion with a uniform velocity makes the tick of the alarm clock slower by a factor of γ. If the spring with the load is orientated along the motion rather than perpendicular to it, the dependence of the mass, as well as elasticity, on the speed of motion v will be somewhat different. The mass will grow by a factor of γ^3 in accordance with the second rule of relativistic dynamics ($m \not\propto \gamma^3$), while the elastic modulus will increase by a factor of γ ($k \not\propto \gamma$). As for the final result, it will be the same – the clock tick (the frequency of the oscillations of the pendulum) will become slower by a factor of γ.[1]

2.3.5. A clock with a suspended pendulum

Let us now consider a clock with a suspended pendulum. Let us, however, place it not on the Earth but in the cabin of a spaceship moving far away from the Earth with the acceleration g_0 equal to its terrestrial value. The direction of acceleration will be regarded as vertical. Everything in this ship goes on in exactly the same way as it does on the Earth. We preferred a spaceship just because it is easier there to ground our reasoning and to better understand the essence of the issue.

---

1 With the longitudinal forces unaffected by motion, the growth of the elasticity by a factor of γ can be explained in the following way. Let the spring at rest be compressed by a certain external force. When such a compressed spring is set in motion, its every part remains in equilibrium, given the external force is kept unchanged. In other words, not only the spring itself, but also its deformation undergoes the Lorentz contraction. Since the elastic modulus is a ratio of the external force to the deformation, the decrease in the deformation leads to the growth of the elasticity by a factor of γ.

2.3. STOPPING TIME

Let us suppose that the space ship which is accelerating in the "vertical" direction is also given a uniform velocity v in a "horizontal" direction. The force of jet thrust is caused by the chemical reactions of combustion, which give rise to a flow of gases thrown "downward". The gases are fired back under the action of interatomic electromagnetic forces arising in the course of the reaction. These forces are directed "vertically", i.e. transversely to the uniform velocity of

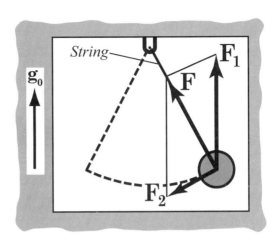

Fig.17 The cabin of a space ship with a pendulum swinging inside it. The ship is being accelerated "upward" with a terrestrial acceleration $\mathbf{g}_0$. The force $\mathbf{F}$ exerted on the load on the part of the string is resolved into the two components: $\mathbf{F}_1$ and $\mathbf{F}_2$, of which the first is accelerating the load together with the cabin, and the second makes the load swing.

the ship. Our previous experience, however, tells us that the transverse forces must become γ times weaker, provided they are registered by stationary instruments. Meanwhile, the mass of the ship grows proportionally to γ. Thus, the "horizontal" motion of the ship with a uniform velocity v results in reduction of the "vertical" acceleration of the ship by a factor of γ^2:

$$g = \frac{g_0}{\gamma^2}.$$

The ship's cabin with a suspended pendulum is shown in Fig.17. The uniform velocity $\mathbf{v}$ points out of the picture, so the plane of swinging is perpendicular to $\mathbf{v}$. On the part of the suspension, the load of the pendulum is acted upon by force $\mathbf{F}$ that can be resolved into two forces: $\mathbf{F}_1$ which makes the load accelerate together with the cabin, and $\mathbf{F}_2$ which is responsible for the oscillations of the pendulum relative to the cabin. When the cabin does not have the velocity v, force $\mathbf{F}_1$ is equal to $m_0 g_0$. Force $\mathbf{F}_1$ is directed "vertically", $\mathbf{F}$

acts along the suspension string, and F_2 is perpendicular to F. Because the directions of all three forces are known, their magnitudes will also become known once the magnitude of at least one of them is found. Since force $F_1 = m_0 g_0$ is known, we can regard the other two forces as also known and determined by F_1.

Let us now see what will happen to the forces when the cabin, accelerating "vertically", acquires also a uniform velocity which is directed "horizontally". According to the first rule of relativistic dynamics, the mass of the load will increase up to the value $m = m_0 \gamma$. The force

$$F_1 = mg = \frac{m_0 \gamma g_0}{\gamma^2} = \frac{m_0 g_0}{\gamma}$$

will become γ times weaker, as was to be expected. Because the directions of all the three forces remain the same, the other two forces will also become γ times weaker. Thus, everything will happen just in the same way as with a transverse spring — the returning force F_2 will become γ times weaker and the mass of the load will be γ times greater. Hence the frequency of oscillations will be γ times lower, which does not differ from the case of a spring-driven clock, as was to be proved.

If the plane of the swinging of the pendulum is directed along the line of motion (vector v lying in the plane of the drawing in Fig.17, and directed either to the right or to the left), then the situation will be more complicated. The force, acting on the load on the part of the string, will make an oblique angle with velocity v, which will bring us into the realm of the third rule of relativistic dynamics. In addition to it, the Lorentz contraction of the string will also to be taken into account inasmuch as the string is inclined to the "vertical" direction. This will somewhat increase the curvature of the load's trajectory. The derivation becomes cumbersome, so we do not give it here. But the result will be the same. The frequency of swinging will be reduced by a factor of γ.

2.3.6. An hour-glass

Let us now take a clock of quite different design – for example an hour-glass. Let us show that even this clock has good reasons to make its rate slower. The hour-glass may consist, for example, of two wide glass tubes set up one on the top of the other and connected through a narrow neck. Through this neck, under the action of gravitation, the sand passes from the upper tube to the lower in a certain period of time. The level of the sand in the lower tube indicates the time that has passed since the hour-glass has been turned over. The faster the sand passes from the upper tube to the lower, the faster the clock's rate. All the mass of the sand in the upper tube is acted upon by the force of gravity, but only a tiny part of it at a time pours down through the narrow neck. The process is similar to that when a large crowd of people tries to pass through a narrow door – everyone tries to get outside, but it is impossible for many people to do it at once. Only a few of them can be getting through the door at a time, and all the others have to wait for them. The rate of their pouring outside at a certain moment of time depends on the swiftness of those two who are passing through the door at that very moment. The same refers to the grains of sand. The lowermost grains within the neck can fall freely. But they cannot fall instantly. The clock rate depends on how fast the falling grains of sand will be vacating the passage for the next layer of the grains. Every freely falling grain will have to cover a certain path S, until it vacates the place for the other grains. The length S itself is of no importance to us. We want only to see how the motion of the hour-glass with a uniform velocity will increase the time it takes one grain to cover the path S. Once again, let us use an accelerating space ship as a test ground for our consideration. We remember that the "horizontal" motion of the ship with a constant velocity **v** results in decreasing its "vertical" acceleration by a factor of γ^2. On board of an accelerating ship, a freely falling grain does not, in fact, take part in the common acceleration, while all the other grains together with the frame of the clock are accelerating upward. Their acceleration is equal to $g = g_0/\gamma^2$, where g_0 is the acceleration of the ship when it does not move "horizontally". Due to acceleration, after a certain time t, the displacement of the departing grain relative to all the other grains will be $S = gt^2/2$, which brings about the time t it takes the departing grain to give the room for the subsequent grains:

$$t = \sqrt{\frac{2S}{g}} = \gamma\sqrt{\frac{2S}{g_0}}.$$

Since the time t has proved to be γ times greater than in the case of no "horizontal" motion, the motion with a constant velocity makes the rate of the hour-glass slower by a factor of γ.

2.3.7. Other examples

The list of clocks could be continued. But it is already too long as it is. If the reader had no time, or was not in position to assimilate all the details, it does not matter. It's more important to grasp the main idea: *However sophisticated the design of a clock might be, there will always be good reasons for slowing its tick down by a factor of* γ. Because this rule is valid regardless of the clock's design, the slowdown of a clock in motion can be noticed only by means of clocks at rest, disposed, say, along the trajectory of motion. It makes no sense to use the clocks in motion for that purpose because all the clocks involved would undergo the same slowdown and none of them would be able to denounce another clock as ticking more slowly. If there were at least a single clock whose tick did not become slower by a factor of γ, it could be immediately used for detecting the absolute motion and even for measuring the velocity of this motion. But, in spite of all efforts, no one, so far, has either found out or designed such a clock, and we have every ground to believe that such clocks do not exist at all.

The universality of the clock's slowdown means that when in motion, every process or reaction − physical, chemical, biological, physiological etc. goes on in a slower tempo. Even the beating of the pulse or the aging of the tissue of a living body becomes slower. Imagine there are two twin brothers, one of them starts on a space travel with a velocity approaching the velocity of light, and the other stays on the Earth. Suppose that ten years later (by terrestrial time) the space traveler returns to the Earth. During that time, his brother who stayed at home became ten years older, while the traveler aged γ times less. If for instance $\gamma = 5$, he became only two years older. So the traveler, after his return, will be eight years younger than his twin brother. Only two years will have passed for him during the time of his travel. Just that time will be counted off by any clock traveling together with him, whether it is the most precise

2.3. STOPPING TIME

chronometer or the pulsation of his heart. In order to return, he will have, though, to retard his motion at some place, and then to accelerate on the way home. Our traveler must experience an acceleration. Otherwise it would have been impossible for him to return home. When accelerating, his clocks (either mechanical or biological) may have another rate, which, strictly speaking, may affect a comparative age of the brothers at their meeting. But it is not essential for our estimations. If acceleration has a noticeable effect on the comparative age of the brothers in our mental experiment, it is always possible to make the main part of the rout longer without changing the start-and-end parts of the voyage where the acceleration is taking place. Then the "rejuvenation" of the traveller, on the main part of the rout, can be made as impressive as you like, so that the change in age on the start-and-end parts of the voyage will become negligible as compared with the ultimate difference in the age. Though the start-and-end acceleration is inescapable, the main part of the "rejuvenation" does occur on the main part of the travel where the velocity is uniform. We will return to this interesting example in Section 2.7.

In 1960, half a century after the birth of relativity, this fact was proved even experimentally. The role of the twins was played by the nuclei of atoms of iron. The nuclei were inside a solid specimen and underwent thermal oscillations. When oscillating the traveling nuclei developed speeds up to 200-300 m/s, that corresponded to the room temperature. The role of the twin who remained at home was played by similar nuclei of another iron sample, cooled off to the temperature of liquid nitrogen (-200^0 C). The iron nuclei emitted γ-rays, that are in fact the same electromagnetic waves as light, but having a much higher frequency. The hot nuclei, vibrating with a higher frequency, played the role of the space traveler, who made a lot of round trips during the time of the experiment. The age of the "travelers" was determined by the number of γ-ray oscillations emitted by them. This number was compared to the number of the oscillations, emitted for the same time by the cold iron nuclei, i.e. by the twin staying at home. The time of observation was 1.5×10^{-7} sec. The experiment proved that, during that time, the twin who stayed at home aged γ times more than his traveling brother. Though by a common standard the time of the experiment was very short, the "twins" had time to experience as much as approximately 5×10^{11} γ-ray oscillations, and the traveling brother was always about half an oscillation younger. The precision of the experiment was sufficient to register the difference in age as small as one tenth fraction of an oscillation. This experiment proved that the "rejuvenation" of the travelers was caused by the uniform velocity, and not by their positive or negative acceleration. Had the acceleration affected the "rejuvenation" of the traveling nuclei by more than 10 %, it would have been registered in the experiment.

The clock tick slowdown was also measured in many other experiments. The most vivid example is that of observing the cosmic rays – streams of particles reaching us from the very depth of the cosmos. In the upper layers of the atmosphere, these rays, colliding with the particles of the air, generate streams of the so-called secondary particles. Many of these secondary particles reach the surface of the Earth and are registered at the mountain observatories. Among secondary particles there are μ mesons (muons) whose lifetime is considerably shorter than the time of their flight from the upper layers of the atmosphere to the earth's surface. Nevertheless they are reliably registered at the surface observatories. The speed of these muons is so close to the speed of light that the tempo of their living experiences 4-fold slowdown. That's the only reason why they succeed in covering the path which, according to the terrestrial clocks, would take as much as about four their lives.

2.3.8. Summary

Our story of moving clocks can be ultimately summarized in the following way:

1. **The motion of any clock with a constant velocity makes its tick slower by a factor of γ. This slowdown can be registered by instruments at rest.**
2. **Because the slowdown of the tick of different clocks moving with the same uniform velocity is the same, the instruments moving together with the clock cannot register this slowdown.**

From the first rule it follows that the time t', measured by a moving clock, is γ times less than the time t, measured by a clock at rest:

$$t' = \frac{t}{\gamma} . \qquad (2.24)$$

The coordinate x of a moving clock varies with time as $x = vt$. If this expression is substituted for the coordinate x into the Lorentz

2.3. STOPPING TIME

transformation (1.13) (see page 48), then the fictitious time t', introduced by Lorentz, will be expressed in terms of the true time t in the following way:

$$t' = \left[t - \frac{vx}{c^2}\right]\gamma = \left[t - \frac{v^2}{c^2}t\right]\gamma = t\left(1 - \beta^2\right)\gamma = \frac{t}{\gamma}.$$

This is, however, nothing else but the equation (2.24). One more very important variable t', introduced by Lorentz, begins to acquire a physical meaning. It turns out to be the time, measured by a clock in motion. One more Lorentz transformation is coming to life. But it is only half-revived as yet. It has acquired a physical meaning only for the case when the equality $x = vt$ is observed, or in other words, when the clock involved is in motion and its readings are compared to those of many stationary clocks which it passes by.

But the transformation (1.13) can also be used in another way. Imagine that an array of clocks is moving in a single file at a constant speed v along the x-axis. At a certain moment of time t, the clocks are located at different points along the x-axis. For the sake of simplicity let us assume that $t = 0$. Having substituted $t = 0$ into the transformation (1.13), we obtain:

$$t' = -\frac{vx}{c^2}\gamma. \qquad (2.25)$$

What can it mean? If variable t' designates the readings of different clocks in motion, which are now at different distances x from the origin of the x-axis, then it follows from (2.25) that now, i.e. at the same moment of time $t = 0$, these clocks show quite different times t', depending on their location – the farther the clock in the direction of motion, the more time is lost by the clock. As for the clocks, located now at the points whose coordinates x are negative, i.e. displaced against the motion from the origin $x = 0$, they, on the contrary, gain time. And only the clock that is at the origin $x = 0$ shows the correct time $t = 0$.

From Part 1 we know of the similar phenomenon, observed in the imaginary world of Lorentz who even introduced a special term *"local time"* and assigned it to t'. Can it be that such extraordinary events take place not only in the imaginary world of Lorentz, but also in our real world? And if it is the case, then what makes the moving clocks behave in such a curious way? Maybe it is somehow connected with the slowdown in their tick? The answers to these and other questions will be given in the next section.

2.4. Time zones on a speeding platform

where we will see that an array of spatially separated clocks which are arranged in a single file and whose hands are synchronized with each other by anything you like (e.g. by light signals) have a good reason to lose or gain time according to their positions in the file as soon as they are set in motion with a constant velocity

2.4.1. Transportation of a clock along a resting platform

In the previous section our study was focused on the slowdown in the tick of a clock under the action of its motion with a uniform velocity **v**. Now, we will approach this problem from somewhat another side. Our goal will be not the tick itself, but rather the effect of the tick variation on the clock's reading. This effect takes place every time when the clock of any design is changing its location. While being moved from one location to another, every clock is ticking more slowly, which makes it slow – at least a little bit – with respect to a similar clock whose location has not changed.

Let us put the following question. How will the reading of a clock change if it is transported with a constant velocity u from point A to point B, a distance l apart? What time will have been lost by the clock by the end of such transportation?

The time it takes the clock to travel between A and B will be $t = l/u$. While being transported, the clock is ticking more slowly by a factor of γ, and if at the start it reads a zero time (let it be a stop-watch triggered at the start of the travel), then at the finish it will show the time $t' = t/\gamma$. This reading differs from the time t shown by a stationary clock (triggered simultaneously with the transported one) by the following value:

$$\Delta t = t' - t = t\sqrt{1 - \frac{u^2}{c^2}} - t \cong$$

$$\cong t\left[1 - \frac{u^2}{2c^2} - 1\right] = -\frac{u^2 t}{2c^2}. \tag{2.26}$$

Speed u was assumed to be small here as compared with the speed of light c in order to justify the use of the approximation (2.13). (See page 103.) If time t is expressed in terms of distance l by means of the substitution $t = l / |u|$ in (2.26), we arrive at the following result for the time lost by the clock in the process of its transportation:

$$\Delta t = t' - t = -\frac{l|u|}{2c^2}. \tag{2.27}$$

To make sure that the transported clock has lost time by just this value, we could mount a TV receiver halfway between A and B and compare the image of the transported clock B on the TV screen with the image of the clock A which stayed at point A during the transportation. The difference in readings between the two images would be equal to (2.27).

Equations (2.26) and (2.27) suggest the following conclusions:

1. The time lost by the clock during its transportation, does not depend on the direction of transportation. The velocity u in the right-hand part of (2.26) is squared, which makes the result indifferent to the sign of u and, hence, to the direction of the transportation.
2. The time lost by the clock is proportional to the magnitude of speed u with which the clock has been transported. If the clock is transported even very far, the time lost by it can be made as small as you like given the speed of the transportation is small enough.

While the first conclusion follows from the slowdown of the clock tick almost automatically (and could have been drawn even without deriving (2.27)), the second conclusion is not so obvious as it might seem on the face of it. The more slowly the clock moves from A to B, the weaker is the slowdown of its tick but the longer is the time of the travel. These two effects act against each other, which could bring us to the situation in which the time lost by the clock during its transportation, would be determined only by the distance l, regardless of the velocity u. But this is not the case. According to equation (2.26), the slowdown of the tick is proportional to u^2 and thus affects the reading of the clock stronger than the increase in the time of its travel $t=l/|u|$, which is inversely proportional to only the first power of u. That's why the substitution of $t=l/|u|$ into equation (2.26) results in formula (2.27), according to which the time lost by the clock depends not only on the distance l between points A and B, but also on the speed u of the transportation. It turns out that in order to "get younger" it is not sufficient to cover a long distance. It is also necessary to do it fast enough.

An inverse assertion is also true: If the velocity u of the transportation is small enough, then the time lost by the clock can be neglected, and we may be pretty sure that the reading of the clock does not noticeably depend on its location. This widens our ability to match the clocks located at different points of space very far from each other.

How do we generally verify our clocks? By wireless time signals, transmitted by radio waves with the speed of light c. On the Earth this method is fair, because the time of propagation of a radio signal round the Earth is as short as about 0.1 sec. In our common practice, this delay is negligible. But suppose we are not on the Earth, but, say, on the Mars. The time it takes a radio signal to cover the distance between the Earth and the Mars is as long as from 3 to 15 minutes. Such a long delay would have to be taken into consideration if we were verifying our clock on the Mars by signals transmitted from the Earth. It is not an easy task to take this delay into account. We must know the exact distance from the Earth to the Mars, that depends on the season of the terrestrial year, and on the season of the Mars's year, and on many other different things. Isn't it much more practical to take a timepiece, verify it on the Earth by means of wireless time signals and then convey it to the Mars. If this were done with the fastest rocket, whose speed of motion was, say, 10 km/sec, the motion would make the clock slow by only 0.01– .02 sec, which is 10 times less than the error that would have been made, had we verified the clock by wireless time signals on the

Earth. Had we taken the timepiece to the Mars, we could compare its readings with the time signals received from the Earth. The difference in time could be used to estimate the distance between the Earth and the Mars and also all the changes in that distance for various positions of the two planets in their orbits. Had the precision 0.01 *sec* occurred insufficient, the timepiece could be specially delivered to the Mars by a slower rocket whose speed would be not more than, say, 100 m/sec. The delivery would have taken dozens of years, but the time lost by the timepiece would be no more than 0.0001 *sec* or so. As for the timepiece itself, we regard it as absolutely stable, modern technology based on lasers and masers being a good reason for it. The possible instability of the modern standard of time does not exceed 10^{-13} or so.

2.4.2. Transportation of a clock along a speeding platform

So far, the test clock, i.e. the clock whose lost time was an object of our investigation, traveled between the points A and B, a distance l apart. The points might be located at any place, so our "proving ground" could be imagined as a huge resting platform or plate. The stakes A and B could be installed at any place of the platform according to our wish. The distance l between them could be measured by a stationary meter stick. The stakes could be equipped with stationary clocks, matched to each other through the wireless time signals with taking into account the delay l/c. Eventually, it has become clear that the matching of the clocks can be performed directly at one of the stakes (say, at stake A) and then one of the clocks could be slowly and carefully conveyed to another stake (say, to stake B). The result turned out to be the same as when using wireless time signals with the correction for the time of their propagation.

Everything that has been said can be ultimately reduced to a simple rule: *If the Platform is at rest, the location of a clock on the platform does not affect its readings.* Does this rule remain in force if the platform is set in motion with a constant velocity v? To make it clear, it is sufficient to answer the following question: what time will be gained or lost by the clock during its transportation with a velocity u from the point A to the point B, a distance l apart, when

2.4. TIME ZONES ON A SPEEDING PLATFORM 129

those points are speeding at a uniform speed v along the straight line AB? The velocity u is counted relative to points A and B. By putting the question in this way, we have set all our "proving ground" in motion with a velocity v, keeping ourselves at rest. Armed to the teeth with stationary measuring instruments (available at any point of space), we use them in order to know everything that is happening on the speeding platform. Both velocities (v and u) are measured with our stationary instruments, and so is the distance l between the moving posts. And the effect of the transportation on the time lost or gained by the clock will be also estimated through such stationary instruments. In other words, when describing the events that occur on a speeding platform, we rely only upon the readings of the instruments at rest and on nothing else. Such an approach seems quite reasonable because the instruments in motion could be "damaged" by their motion and hence regarded as unreliable.

Even without any derivations, it can be figured out that the motion of the platform will substantially affect the behavior of the clock which is being transported on a speeding platform.[1] First of all, it strikes the eye that now the effect of the transportation on the readings of the clock must depend on the direction of the transportation. Let the velocity of the platform v be directed, for example, from point A to point B. Then during the transportation of the clock from A to B the net velocity of its motion will be higher than in the case of a clock fixed to the platform. Thus, the transportation of the clock from A to B must make this clock lose some time as compared with a similar clock left at point A. On the way back, on the contrary, the net velocity of the clock will be lower than the velocity of the platform, and, thus, the transportation of the clock will make the clock gain time. This means that formula (2.26) cannot be applied to the case of a speeding platform. According to that formula everything must be quite symmetrical, so that any transportation of the clock on the platform would make the clock only lose time, and the lost time would vanish if the speed of transportation u approached zero. The sign of u, which is squared there, would not then affect the time Δt lost by the clock. But in the case of a speeding platform the sign of u does affect the sign of the time Δt which is lost or gained by the clock and whose dependence on the velocity u proves to be quite different. Therefore, from now on, instead of the distance l, we will use

---

1 Please, double your attention here! The platform with the stakes A and B has begun to move. It is the relativity of simultaneity – the heart of special relativity – that we are approaching very closely.

the change of the coordinate of the clock $\Delta x = \pm l$, that will be assumed positive if the clock is transported from stake A in the direction of the motion, and negative – in the opposite case.

Until the clock is fixed to the speeding platform, its tick is slower by a factor of γ. But once its transportation begins, the factor of the slowdown of its tick becomes equal to γ_u, defined by the speed $v + u$:

$$\gamma = \frac{1}{\sqrt{1 - \frac{v^2}{c^2}}}; \qquad \gamma_u = \frac{1}{\sqrt{1 - \frac{(v+u)^2}{c^2}}}. \qquad (2.28)$$

Let us use t to designate the duration of the clock transportation. This time will be needed by the clock to cover the distance Δx between A and B: $\Delta x = ut$. Quantities Δx and u may be either positive or negative depending on the direction of the transportation: whether it is done along the platform's motion or against it. If the clock were not transported, its hand would displace by t/γ points because in that case it would move together with the platform. The same clock, while being transported, will be moving either faster than the platform (speed u is then positive), or more slowly than the platform (the speed u is negative), and its hands will advance by t/γ_u points. Thus, the following difference in readings Δt will arise between the clock transported from A to B and the clock stationed at point A:

$$\Delta t = \frac{t}{\gamma_u} - \frac{t}{\gamma} = t\left[\sqrt{1 - \frac{(v+u)^2}{c^2}} - \frac{1}{\gamma}\right] =$$

$$= t\left[\sqrt{1 - \beta^2 - \frac{2uv}{c^2} - \frac{u^2}{c^2}} - \frac{1}{\gamma}\right] =$$

$$= \frac{t}{\gamma}\left[\sqrt{1 - \frac{2uv\gamma^2}{c^2} - \frac{u^2\gamma^2}{c^2}} - 1\right]. \qquad (2.29)$$

2.4. TIME ZONES ON A SPEEDING PLATFORM

If the speed u of the clock transportation is close to the speed of light c, the influence of u upon Δt is of a rather intricate character. But if u is small enough as compared with c (this case being of the main interest to us), then expression (2.29) can be simplified to a great extent after dropping the item $(u^2\gamma^2)/c^2$ and escaping the square root by means of the approximation (2.13) on page 103:

$$\Delta t \cong \frac{t}{\gamma}\left[\sqrt{1-\frac{2uv\gamma^2}{c^2}}-1\right] \cong \frac{t}{\gamma}\left[1-\frac{uv\gamma^2}{c^2}-1\right] = \\ = -\frac{tuv\gamma}{c^2} \quad (2.30)$$

Unlike formula (2.26), the change in the readings of the clock Δt proved proportional not to the second, but to the first power of velocity u. If we make a transition from the time of the transportation $t = (\Delta x)/u$ to the distance Δx between A and B, velocity u will disappear altogether:

$$\Delta t = -\frac{v\gamma\Delta x}{c^2}. \quad (2.31)$$

We have arrived at the following remarkable result:

A slow transportation of a clock along a speeding platform, either in the direction of the motion or against it, makes the clock lose time or gain it by the value Δt that depends on the distance Δx of clock transportation and is independent of the velocity u of that transportation.

Independence of the time Δt, lost or gained by the clock, from the velocity u of the transportation has a rather simple explanation: the smaller is the velocity u, the less will be the change in the tick of the clock during the transportation, but the longer will be the time taken by the transportation. These two effects balance each other, and eventually it turns out that the time lost by the clock or gained by it is determined by distance Δx alone and does not depend at all on velocity u. The positive displacement Δx of the clock always brings about a negative value for Δt, i.e. the clock loses time, while a negative Δx causes the clock to gain time.

If point A is assumed to be at the origin of the frame of reference ($x = 0$), then $\Delta x = x$, and regularity (2.31) takes the form that must be familiar to us from previous sections:

$$\Delta t = -\frac{vx\gamma}{c^2}. \qquad (2.32)$$

So, moving slowly along the x-axis on a fast platform, we always pass from one time zone to another, the hands of our clocks shifting back or forth just by themselves. "The hands of our clocks" should be understood in the widest sense of the word. Not only the hands of the clock are slow or fast, but also the beating of the heart, the sound of music, the flight and singing of birds, the work of motors, machines, and other devices, etc.

2.4.3. Origination of time zones on a speeding platform

Let us suppose that you are flying from Vladivostok to Moscow with a speed close to the speed of the rotation of the Earth. While you are flying, the sun is hanging almost stopped over the horizon, and so is the local time, if counted according to the sun's position. Starting from Vladivostok at noon, by local time, and having been in the air nine hours, you arrive in Moscow at 2 p.m. by the Moscow time, so you have to reset your watch seven hours back. Now imagine that the hands of your watch are shifting back just by themselves, all the events on board the plane are going in a slower tempo, and in the course of the flight, you have become older not by nine but by two hours only. Things would be just like that if Vladivostok and Moscow were placed on the speeding platform whose speed was close to that of light $v/c = 0.9999995$.

On arriving in Moscow-on-the-Platform you have become seven hours younger than your twin-brother who stayed in Vladivostok-on-the-Platform. Does it mean that on returning to Vladivostok by the same plane you will be 14 hours younger than your twin? Oh no, it does not. On the way back, the value x in the expressions (2.32) will become negative, and Δt – positive. That means that on the way back everything will be the other way around. This time the hands of the clock will run ahead, all the processes on board the plane will be going on faster (the way it is after a slackened filming), and after a nine-hour flight you will become older by as much as 16 hours, i.e. by seven hours more than if you had stayed in Moscow. The "rejuvenation", acquired by you when flying to Moscow, will be lost absolutely when you return to Vladivostok. There will be no difference in age between you and your twin brother, who will meet you in Vladivostok. Your watch will show the same time as his, as if you had not flown to Moscow and had been all that time with him in Vladivostok. As for your own impressions, you will not be able to feel either your "getting younger" on the way there, or your "aging" on the way back, because everyone and everything around you will be "getting younger" on the way there and "aging" on the way back, same as you. Whatever you looked at, whatever instruments you took, you would never be able to feel or learn that you were getting younger or older, if every instrument was moving together with you. But the instruments at rest, which were used by us to track down your adventures,[1] would register both the retardation and acceleration of all the processes on board the plane on the way forth and back. So you would be able to learn about your "getting younger" and "aging" by means of addressing us by radio and inquiring about the readings of our stationary instruments. Our account of your conversions would cause a lot of mistrust on your part, because you yourself did not notice anything of the kind. [2]

So, the conversions, you have experienced in the air, proved unnoticeable to you. A question arises. Could they be noticed by the residents of your speeding platform, by those who did not take part in your travel and watched you, say, by TV from Vladivostok, or, still better, from the point halfway between Vladivostok and Moscow? Such a choice of an observation post would allow to make an unbiased comparison between the readings of the clock that remained in Vladivostok and the readings of the clock that was brought to Moscow and lost as much as seven hours. Due to the midway position of such an observation post, it is possible to do without caring of the

---

1 Nothing would change if these instruments were assumed to be fixed to the ether, that will fade away in the next two sections.
2 This reasoning is valid provided your displacement along a speeding platform is sufficiently slow, i.e. with a speed much lower than the speed of light. If you are on board the plane, this condition is satisfied well. As for the platform, it must be rushing at full speed, close to that of light.

times which it would take the radio signal from Vladivostok and that from Moscow to arrive at the screens of the two TV sets placed side by side to each other. Whatever those times might be, they would be the same from the standpoint of the residents of the platform. The paths of the signals being equal, and the residents not aware of the platform being in motion, they would reason like this:

"Though there is a certain delay before we see each clock on the screen, this delay must be absolutely the same for each clock. So if the clock in Moscow is seven hours slow in comparison with the clock in Vladivostok, the image of the Moscow clock will be also seven hours slow than the image of the Vladivostok clock." But the residents of the speeding platform are unaware of their being in motion and that "in fact" (i.e. judging by the readings of the instruments that are at rest) the midway post moves with the speed v against the radio signal from Moscow and is running away with the same speed v from the radio signal coming from Vladivostok. That's why "in fact" the TV image of the Moscow clock will arrive at the observation post earlier, and that of the Vladivostok clock − later than it is believed by the residents of the platform. If l is the distance between Vladivostok and Moscow (both of them on the platform), then the time it will take the image of the Moscow clock to arrive at the screen will be $l/(2[c+v])$, while the time taken by the image of the Vladivostok clock to reach its destination will be $l/(2[c-v])$. Within these time intervals, as measured by clocks at rest, the hands of the Moscow clock will shift by $l/(2\gamma[c+v])$ points, and the hands of the Vladivostok clock − by $l/(2\gamma[c-v])$ points, with taking into account that any clock fixed to the speeding platform ticks γ times more slowly. Thus, the images of the two clocks, simultaneously registered at the midway observation post, are delayed by the time $l/(2\gamma[c+v])$ for the Moscow clock, and by the time $l/(2\gamma[c-v])$ for the Vladivostok clock. The difference between these two delays may be regarded as an error of observation, caused by the fact that the residents of the platform are unaware of their being in motion. Let us calculate that difference. It is equal to

2.4. TIME ZONES ON A SPEEDING PLATFORM   135

$$\Delta t = \frac{l}{2\gamma[c-v]} - \frac{l}{2\gamma[c+v]} = \frac{l[c+v-c+v]}{2\gamma[c^2-v^2]} =$$

$$= \frac{vl\gamma^2}{\gamma c^2} = \frac{vl\gamma}{c^2}.$$

Take a good look now at the expression (2.31) that determines what the residents of the platform wanted to detect on the TV screens. The observational error is of the same magnitude as what they wanted to measure and has the opposite sign. This means that the images of the Moscow and Vladivostok clocks on the TV screens will show the same time. It is not for the first time that we are face-to-face with such a phenomenon. The time lost by the clock due to its transportation, and the difference in time of the propagation of TV signals (though, on the face of it, the two things seem to have nothing in common) are acting in unison here, as if they have conspired to conceal the motion of the platform from its residents. If it were not so and the readings of the clocks on the TV screens were just a bit different from each other, that "bit" could have been immediately made use of by the residents to determine the velocity of their absolute motion, or, which is the same, the velocity of the ether drift.

Thus, without looking out of the limits of the moving world, no one among the residents of that world is able to learn that the transfer of the clock from Vladivostok to Moscow made it seven hours slow. But at the same time, the instruments at rest. which can be thought of as fixed to the ether, persist in repeating one and the same thing:

"The clock is slow, the clock is slow... Its being slow is caused by the transportation. It shows now a wrong time! The residents of the moving world are absolutely helpless! What their TV screens show is shifted in time, so as to conceal from them the clock's delay. The residents are unaware of the curious things happening in their world. And only we – the inhabitants of the ether – can see everything in the true light."

But never mind, in Section 2.6 we'll see how the instruments fixed to the ether will be compelled to change their opinion about which of the instruments are right and which of them are wrong.

Meanwhile we will take a look at our mental experiment with the TV sets and clocks from another point of view. Suppose that during your flight from Vladivostok to Moscow (both on the platform) you decided to wind up your watch and by mistake you set it wrong. You have no other clock by you, and the time is gone. What can you do? Must you return to Vladivostok to set the watch there by the local time standard and then transport your watch to Moscow again? The experiment with TV shows that you can do without it. On landing, you bring your wrong watch to the TV center and get connected with the midway observation post. You learn the time shown at the moment by the image of the Vladivostok clock and move the hands of your clock until the images of the two clocks (yours and that from Vladivostok) on the two TV screens at the midway post begin showing the same time. After that, you may be sure your watch is OK, it shows the correct time, as if nothing had been wrong with it. It is this method of matching the clocks located in distant points of space far from each other (*spatially separated clocks*) that was proposed by Einstein in special relativity. There being no TV in those days, he suggested synchronizing the clocks by means of two light signals simultaneously sent in opposite directions from the point located halfway between the two clocks that are to be matched. The result is sure to be the same. Light signals used by Einstein are known now as wireless time signals. When matching the clocks, such signals can be sent from any location, provided a correction is made for the difference in distances from the clocks being matched to the sending station. All these methods of matching the clocks give the same result as a slow transportation of each clock from the time standard to the clock's destination.

Thus, in the moving world, the readings of the clocks, matched to each other, depend on their location. This world is divided into time zones gradually changing into each other along the direction of motion. When passing from one zone to another, the hands of the clock shift automatically, just on their own, as though the clocks were continuously matched through light signals. This means, for example, that simultaneity or non-simultaneity of events happening at different places of the moving world depends on whether the clocks by which these events were registered belong to the moving or to the stationary world fixed to the ether. The events, that are simultaneous according to the stationary clocks, prove to be non-simultaneous according to the clocks in motion, and the other way around. When one learns about it for the first time, one wonders if there is a

2.4. TIME ZONES ON A SPEEDING PLATFORM

way of putting all these tricks in order. It seems by intuition that absolute simultaneity is comprehensible just by itself — it is something that goes without saying. What makes us think so? The reason is quite simple. We take it obvious that there are some means of instant transmitting any information from one place to another. We do not care too much how to perform such an instant transmitting in practice, but somehow believe that the way of doing so must exist. Actually, no such way is known to science. As long as we know, the speed of light is the ultimate limit on transmitting any information, and we do not see even a hint of a possibility of overcoming the light barrier.

What would happen if instant signals existed in nature?

It would be quite different if instant signals existed in nature. Then all the clocks on a speeding platform, whatever place they are, could be instantly matched so as to show the same time, which would not differ from the time shown by the clocks at rest. This would greatly simplify the situation, wouldn't it? But do not jump to conclusions! It is not so obvious as it might seem from the first sight.

> Let us give way to our imagination. Because it is next to impossible not only to consider signals whose velocity might exceed the velocity of light, but even to think about them, we have to transfer ourselves inside a fiction novel (everything is permitted there) and imagine that, some time in 28th century, a great physicist Smartman, discovered a superlight signal, whose speed is greater than that of light. Suppose he got the electron accelerated so fast that its mass had no time to follow the increase of its velocity. Smartman made use of this temporary failure of the electron and pushed it beyond the light barrier. There, the electron faced enormous forces that tried to pull it back to our before-the-barrier world. But Smartman invented an ingenious method of eliminating these forces and got this super-electron accelerated up to the velocity 10 times greater than the speed of light, thus supplying humanity with a means of practically instant communication. Making use of that effect, Smartman at last detected the ether drift and measured its velocity. Having thus discovered the absolute time, he proposed to use it throughout the moving world as the only appropriate method of keeping time. The Central Council on the Platform issued a decree according to which the local

time was forbidden and all citizens on the speeding platform were prescribed to use only the absolute time transmitted instantly everywhere from a single center by means of Smartman's superelectrons.

At first all the citizens enjoyed this innovation very much. "At last the residents of other worlds will stop their gossip about our troubles with simultaneity." Everybody was obeying the decree in the most religious way. But that proved to be not so easy as it had been expected, especially for the people who frequently changed their location. It was very bothersome for them to reset their clocks after every move. Their clocks did not know anything either of the decree or superelectron, and continued to change their readings after every travel in accordance with expression (2.32). Otherwise life on the platform also went on as usual, and soon there appeared people who after traveling too much found themselves in confusion about their own age. The platform moving very fast, the difference between the local time and the absolute time in some remote places was as large as many years. Trying to determine their own age, these people began to study Einstein's relativity very diligently. Coming across equation (2.32), they learned that, wherever they had traveled, their actual age was determined by the coordinate x. That was the clue to restoring their lost age. It would be sufficient to visit their home town and ask their schoolmates how old they were. That would be just their age.[1] Urged by their bitter experience, the travelers-on-the-platform stopped shifting the hands of the watches after every move and returned to the local time. The government soon also realized the inconvenience of its decree and canceled the absolute time, sparing it only for those physical experiments where superelectrons made a significant contribution. Because Smartman was the only scientist who was carrying out such experiments, he eventually proved to be the only person who lived by absolute time. But he did not suffer too much, because he was a well-known stay-at-home.

---

1   To keep the information of their own age was not an easy matter even for those schoolmates who had never left their birthplace. They had to keep a double calendar: one for the government, where the absolute time was registered in accordance with the decree, and the other for themselves, where an entry was made secretly every evening in spite of the government's orders. The time difference between the two calendars reflected the history of the ether drift.

We have told you this incredible story just to demonstrate the utmost firmness of special relativity, which we are approaching very close now. Even if an impossible thing were to happen in the future: a superlight signal would be generated (though nowadays we have no grounds for such a prediction), relativistic effects would continue to take place as if nothing had happened, and special relativity would still embrace all the branches of science and technology known to us today. It is proved by the fact that special relativity, having turned upside down our ideas of space and time, left untouched all the fundamental laws of nature that were known at the moment of its creation.[1] Instead of replacing the laws with other ones, special relativity uncovered in the existing theories many unexpected things, that had never been suspected of even by the founders of those theories or their numerous followers. That is why special relativity cannot be separated from the rest of physics or be opposed to it in any way.

2.4.4. A device for envisioning the relativity of simultaneity

The previous sections tell us that a rod, set from rest to motion with a speed v, is sure to contract its length by a factor of γ. The behavior of a clock, that retards its tick, is similar. And what will be the behavior of two clocks, a distance x apart, that simultaneously start from rest to motion with constant velocity? Some surprises await us there. First let us consider the behavior of these clocks when they are connected to each other through a continuous synchronizing electromagnetic signal. Let the clock A in Fig.18 be a master clock. It has its own mechanism, driving its hands. A continuous electromagnetic wave, emitted from A, is traveling to the second clock B with the speed c. The crests of the wave are indicated in the figure by bold strokes. Let every wave crest, sent from A, correspond to a one-point move of the hand of A. Clock B has not got any driving mechanism of its own. It is a slave-clock that works only by the orders received from clock A through the synchronizing wave. Each wave crest makes clock B shift its hand by one point. Initially clocks A and B are at rest and matched by any of the methods described

---

[1] Except Newton's law of gravitation, which is out of the scope of special relativity. As for the process of setting the bodies in motion, it is thought of in non-postulated relativity as taking place under the action of any forces but gravitational.

above, so as to show the same time. They are fixed to a common solid plate, a distance x apart.

Let us see now what will happen if the whole system, comprising the plate and the two clocks, is suddenly accelerated to the left up to a uniform velocity v, after which it will be moving with a constant velocity indefinitely long. The system can be accelerated by a compressed spring, or by compressed air, or, say, by a jet engine – by whatever you like except the force of gravity. We will discuss gravitational forces later, in Section 2.9, and until then we will abstain from using them. Let us assume that during the acceleration the plate and the clocks remain safe and sound. Then the distance between the clocks will be reduced by a factor of γ. First clock A, and then clock B will slow down their tick by a factor of γ. But at this stage of the explanation, we will turn the blind eye to both these effects, for we have to deal with things much more important than that.

To understand these more important things better, let us make some substitutions. You will act the role of the clock B. For this purpose you settle down in the town of Vladivostok on the Earth. The role of the clock A will be acted by a friend of yours who will settle down in Moscow. The role of the wave crests will be played by the letters that your friend will send you daily by a mail van of a passenger train bringing you just one letter a day. These letters will arrive with an eight-day delay, eight days being the time it takes the Moscow train to get to Vladivostok. Suppose on a fine summer day both of you simultaneously went bicycling West with the same velocity. Suppose it was on your birthday, so that the last letter before the start was sent by your friend from Moscow in a decorated envelope. While bicycling, your friend kept sending you one letter a day. And you continued receiving the letters at the intermediate stations between Moscow and Vladivostok.

Your distance to Moscow is getting shorter every day, and so is the time it takes the train to bring you a letter that had been sent from Moscow before your friend started bicycling. Now you are receiving his letters more frequently than once a day. This will last until you get his letter of congratulation. Until then, you will receive more letters from him (say, two letters more), than you would have gotten had you stayed in Vladivostok. But after you get the letter of congratulation, the rate of receiving the letters will be reverted down and you will be again getting exactly

2.4. TIME ZONES ON A SPEEDING PLATFORM

one letter a day. Now not only the receiver of the letters is coming closer to Moscow, but also the sender of the letters is getting farther from Moscow while moving off in the same direction with the same speed. No matter how far you go, the total number of the letters received by you will always be two letters more than it would have been, had you not gone bicycling. Now imagine the following: every time your friend drops a letter for you, he shifts the hand of his clock by one point. And every time you get his letter, you do the same. (The rest of the time the hands of the clocks do not change their position.) At the first stage of your travel your clock keeps ticking faster. (You receive the letters more frequently than they have been sent.) And after the moment you receive the letter of congratulation, your clock will always be two points fast. This can be explained by the fact that when you went bicycling, some letters were already on their way, and they were reaching you more frequently than all the subsequent letters. And it was due to these letters, that your clock became two points faster than that of your friend's.

Now it is clearer what will happen to the clocks shown in Fig.18 after they are suddenly set in motion. The clocks and the plate are the only things accelerated there. The wave crests being on their way will never experience any acceleration and will continue to propagate freely from the left to the right with the same speed c. Clock B will move against them with speed v and will be affected by them more frequently until it receives the "crest of congratulation" – in Fig.18, it is encircled. Everything what is of interest to us will take place until that very moment, which we will name the end of the transient. Nothing interesting will happen after that moment. The frequency with which the crests arrive at clock B will reduce and become the same as before the acceleration. (As for the effect of the motion on the tick of the clocks, we have agreed not to take it into consideration at least for the time being.)

Let us now calculate how fast clock B will become in comparison with clock A as a result of the procedure described above. For now, we will continue to confine ourselves to the case of a small speed v, so that β^2 could be neglected as compared with β. We can indulge in doing so because the expected effect, as defined by formula (2.32), belongs to the so-called first order effects that are proportional to the first power of β. These effects have to be distinguished from the second-order effects such as the Lorentz length contraction or clock-tick

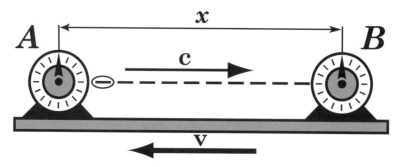

Fig.18. Two clocks A and B are fixed to the plate, which is instantly set in motion from the right to the left with a constant speed v, close to the speed of light c. The clocks are continuously synchronized to each other by an electromagnetic wave. The crests of this wave are shown with bold strokes, propagating from the left to the right. This wave was established long before the acceleration of the clocks. The clocks are shown immediately after acceleration, so that the encircled crest is the last crest emitted before the clocks were set in motion.

slowdown, which are proportional to the factor γ that differs from unity by a value of the order of β^2.

First, we will calculate how many crests there are on the way just before the acceleration occurs. The time it takes the leftmost crest to cover distance x is equal to x/c. Let T be the time it takes the wave to cover the distance between two neighboring crests. Then the time $nT = x/c$ (where n is the desired number of the wave crests) will be taken by the leftmost crest to cover the distance x. If T is regarded as a unit of time (i.e. we agree that every act of sending or receiving a wave crest is followed by shifting the hand of the relevant clock by one point), then the time interval between two successive crests becomes $T = 1$, and the desired number of the crests n is equal just to x/c It is this number of the crests that will be registered by clock B by the end of the transient, when the "crest of congratulation" reaches clock B. Thus, the reading t_B of this clock at that very moment will be equal to

$$t_B = \frac{x}{c} \; ; \qquad (2.33)$$

(both the clocks are assumed to be set at zero just before the acceleration begins.) But what time will be shown by clock A at the same

2.4. TIME ZONES ON A SPEEDING PLATFORM

moment of time (i.e. at the end of the transient)? Because clock A works independently, so that nothing happens to it, it will just show the time needed for the transient to come to its end, i.e. needed by clock B to receive the "crest of congratulation". This crest is moving to the right with the speed c. Clock B is rushing against it with speed v. The initial distance between them is x. Thus, the time it will take them to meet is equal to

$$t_A = \frac{x}{c+v} = \frac{x}{c[1+\beta]} \ . \tag{2.34}$$

Comparing the expressions (2.33) and (2.34), it is easy to see that, at the end of the transient, clock B will show a later time than clock A, i.e. B will be fast as compared with A, or, which is the same, A will be slow as compared with B. What time will be lost by the clock A with respect to the clock B? Now we are ready to answer this question. To arrive at the result it is enough to subtract (2.33) from (2.34):

$$\begin{aligned}\Delta t = t_A - t_B &= -\frac{x}{c}\left[1 - \frac{1}{1+\beta}\right] \cong \\ &\cong -\frac{x}{c}\left[1 - (1-\beta)\right] = -\frac{vx}{c^2} \ ,\end{aligned} \tag{2.35}$$

where we have made use of the second approximation (2.13) (see page 103), dropping thus the addend of the order of β^2. Let us compare this result with formula (2.32). There is no difference between them except the factor γ, that differs from unity by an addend of the order of β^2, which is within our first-order approximation. So, the clocks A and B, after they are set in motion with a constant velocity, begin to show different times, or, in other words, they behave in the same way as the clocks which have always been on the speeding platform. If the moving clocks A and B were observed from the moving midway TV post, their TV images would always show the same time. If these moving clocks are slowly and carefully brought together, they will also show the same time. If they are taken apart again, they will again show different times. In short, there will be no difference between our clocks and the clocks which have been on the speeding platform before the acceleration, or which even were born there. It should be emphasized that, in the process of their resettling to the speeding platform, their hands were shifted just by themselves in accordance with the time zone they happened to arrive at. Now we see of course that actually it was not by themselves that they were readjusted. It was the wave crests on their way at the moment of the resetting that were responsible for everything.

Now, that we have understood the main reason for the change in readings between the two synchronized clocks in the process of their acceleration, we can indulge the lovers of subtleties in improving the accuracy of our derivations through taking into account two more effects neglected above.[1] The first of them is the Lorentz contraction of the clocks' separation by a factor of γ, and the second one is the slowdown of the clock's A tick by the same factor. Let us see how this affects the expressions (2.33) and (2.34). Let x designate now the distance between the clocks A and B after the transient is over and the plate under the clocks has undergone the Lorentz contraction by a factor of γ. Such definition for x makes sense because it matches the meaning of x in the formula (2.32), with which the result of our current derivation will be eventually compared. And what was meant by x, when formula (2.33) was under derivation and we were engaged in counting the number of wave crests between the clocks A and B? Then the distance x designated the space between the two clocks before they were accelerated. Therefore, this time, in (2.33), instead of x we must use the quantity which is γ times larger:

$$t_B = \frac{\gamma x}{c}. \tag{2.36}$$

Quite a different role was played by x in formula (2.34). There, x stood for the *initial* distance between the clock B and the "crest of congratulation " – that very distance which the clock and the crest must cover by joint efforts in order to meet each other in the time t_A. The contraction of the plate will make them meet sooner. (The plate is supposed to end its contraction before the meeting because otherwise our derivations would be too long and cumbersome). The

---

[1] If the reader is ready to believe in the correctness of equation (2.35) with its right-hand side, multiplied additionally by γ, and has no dispose to plunge into subtleties, it is possible to omit the forthcoming non-trivial derivation and jump straight to formula (2.38).

2.4. TIME ZONES ON A SPEEDING PLATFORM 145

marked wave crest will be propagating to the right with the speed c, clock B will be flying against it with the speed v, and, in addition to it, the plate under the clocks will be undergoing the Lorentz contraction and thus enhancing the clocks' movement toward the place of meeting. The latter effect can be formally taken into account if x in expression (2.34) is regarded as a distance between the clocks shortened by the Lorentz contraction. Because we have agreed that x denotes just this contracted distance, we needn't do anything to x in equation (2.34). We have to leave x as it is and to bear in mind that such an absence of any actions automatically implies the effect of the Lorentz contraction upon the duration of the transient.

There remains now only to take into consideration the slowdown of the tick of clock A after it was set in motion. The slowdown of the tick of clock B needn't be considered separately, because that clock is entirely under the control of clock A and of the propagating wave crests. Its readings are determined only by the number of the wave crests, arrived at B, and by nothing else. Not so for clock A. It is a master-clock and we must take into account the slowdown of its ticking since the moment of its acceleration. What effect will it have on our derivation? It will not by any means tell on the expression (2.36) because that expression is responsible for the number of wave crests between the clocks A and B at the beginning of the transient, and this number has nothing to do with the acceleration of clock A.[1] Neither can it affect the time of the meeting of clock B with the marked wave crest, because that crest had been sent before clock A started its motion. So, equation (2.34) is still valid, giving the time of the end of the transient as shown by the clocks at rest. But this time is not coincident now with the reading of clock A, because now clock A is ticking more slowly, and its hand will shift a smaller number of points (by a factor of γ), than it follows from (2.34). Because the left part of (2.34) is equal to t_A, the right-hand part of that equation should be divided by γ:

---

[1] If the acceleration was carried out under the action of gravitation, then the crests of the electromagnetic wave would be accelerated too, which would bring us to another result. But that would be against our agreement which permits us to use any accelerating force except gravitational.

$$t_A = \frac{x}{c\gamma[1+\beta]}. \qquad (2.37)$$

Now, to obtain the difference in the readings of the two clocks Δt, acquired by the end of the transient, it is sufficient to subtract (2.36) from (2.37):

$$\Delta t = t_A - t_B = \frac{x}{c\gamma[1+\beta]} - \frac{\gamma x}{c} =$$

$$= -\frac{\gamma x}{c}\left[1 - \frac{1-\beta^2}{1+\beta}\right] = -\frac{\gamma x}{c}\left[1-(1-\beta)\right] = -\frac{\beta\gamma x}{c} = -\frac{vx\gamma}{c^2}.$$

$$(2.38)$$

While obtaining this formula, we have neither neglected anything nor used any approximations. The result is on the face of it – (2.38) is an exact copy of equation (2.32) rather than an approximation to it.

And yet, our refined derivation involves one arbitrary assumption, and a thoughtful reader might have noticed it. When the Lorentz contraction of the plate was being taken into account, clock B was implied to be displaced toward clock A as a result of the contraction, and not the other way around. The opposite case (with clock A approaching clock B) could also be considered, but it would go beyond the limits of our bicycle model. Other equations would be needed and other reasoning, but the result would be the same. Clock A would be slow relative to clock B by the value, determined by expression (2.38).

The presumptions of the problem might be changed too. Let the clocks be accelerated not instantly, but gradually. (There are no instantaneous accelerations in nature). But the final result would be the same again. And the reason for the difference in the clock readings would be also the same: the wave crests being still on their way. But these crests would work not at once, the way they did when the acceleration was instantaneous, but gradually, which would make it much more difficult to see and analyze the reason for the clocks' behavior. That's why hereafter, for the sake of simplicity, we will often assume that all accelerations happen instantly and without any instant influence on the object of acceleration. All the particles of

2.4. TIME ZONES ON A SPEEDING PLATFORM 147

the object's body as well as their disposition inside the body will be assumed unchanged at the first moment after the acceleration, and only later on a transient will take place resulting in the Lorentz contraction, the clock tick slowdown and the difference in time lost or gained by the spatially separated clocks. The time of the transient will depend on the longitudinal size of the object or the distance between different objects synchronized with each other.

Now we know that the relative shift in the readings of the two accelerated clocks is caused by a firm connection between them through a light wave. The light wave may be replaced by a sound wave or by an electric signal transmitted by wire, or by something else. The result will be always the same. Clock A will be slow by a value determined by formula (2.32). The crests of the electromagnetic wave which are on their way will continue to do their work. But they will do it invisibly, underhand, shielding themselves with other phenomena, seemingly very important, but in fact absolutely irrelevant. Even if the hands of the clocks in Fig.18 were nailed to a solid beam, after fast acceleration, the beam would oddly bend, so that clock A would lose time with respect to clock B by the same value. When the hand of clock A begins pushing the beam so as to impart its motion to the hand of clock B, the latter will begin to feel this push not at once, but some time later. If the left end of the beam displaces together with the hand of the clock, say, downward, then the right part of the beam will learn about it not immediately; and until then it will of course remain indifferent. In other words, the beam will bend. With velocities close to the velocity of light, the most rigid bodies become as soft as rubber, if observed by the instruments fixed to the ether. It seems very strange that the moving observers do not notice these wonders.

What does happen when two connected clocks become independent of each other?

Let us consider now the behavior of the clocks A and B, isolated from each other. Let every clock have a driving mechanism of its own. This time there are no wave crests on the way that could cause the difference between the clocks' readings. What will then happen to the independent clocks if they are suddenly accelerated? Nothing will happen. Why? Just because there are no reasons for any relative changes. The reasons, associated with the dynamics of the acceleration − overload, etc. are not taken into account for the sake of clarity. We have already seen that the main reasons for affecting the clocks were associated not with the process of acceleration as such, but with its consequences, caused by phenomena taking place not within the clocks, but outside them. They were caused only by synchronizing signals, that are absent now. Therefore, the acceleration of the clocks will be performed now without any consequences. If even the acceleration could somehow affect the readings of the two clocks, it would affect them identically. The clocks A and B are accelerated independently of each other and under absolutely the same conditions. Thus, after acceleration they must show the same time. But if these accelerated clocks are put onto the speeding platform, they will feel there as outsiders who do not belong to the harmonious family of the moving clocks, matched to each other. They will be the only pair of clocks showing the same time in different time zones. If they are observed at the midway TV station, their images on the screens will show different times. The difference in time will again be determined by equation (2.32). If the clocks A and B on the speeding platform are slowly brought together, their readings will differ by the same value. To make them suitable for time measurements on the speeding platform, they must be matched to each other either by TV or through a slow transportation with their hands reset. Then they will show different times in different time zones and will no longer differ from any aboriginal pair of clocks in the moving world.

The example with two independent clocks is very instructive. It shows that not all bodies or systems obey the Lorentz transformations when their velocities are changed. Assume that two rods are placed one after the other along the same straight line. If set in motion along this line, their lengths reduce. And what about the distance between the end of one rod and the beginning of the next one? It can undergo any change depending on the conditions of the acceleration. If the rods are fixed rigidly, say, to a table, then, because the table is contracted by a factor of γ, so will the distance between the rods.

But let us take an opposite case. Assume that the two rods are hovering indifferently one after the other under conditions of weightlessness. Let them be electrically charged. (We will neglect their interaction.) Then, switching on an electric field, we can accelerate the rods up to a velocity v. Each of the rods will become γ times shorter, but the distance between their centers of mass will remain the same. Because they are accelerating under absolutely the same

2.4. TIME ZONES ON A SPEEDING PLATFORM 149

conditions. Our rods represent not one, but two separate equilibrium systems. The same refers to clocks. If coupled continuously through an electromagnetic signal, two spatially separated clocks form a united system, just like the two halves of a solid rod. The controlled clock cannot live and act independently of the master-clock. Therefore, when being transferred onto the speeding platform, the slave-clock gets adapted to the master-clock according to the rules and laws of the world in motion. And in conformity with those laws, when in different time zones, the clocks must show different times. But independent clocks do not obey these laws. Nothing links them to each other. Though either of the two clocks, obeying the laws of the moving world, begins going more slowly, their relative readings do not change, and even when they are in different time zones, they keep showing the same time.

2.4.5. Summary

The most difficult and important section of the book is now coming to the end. It's time for us to sum up the knowledge gained by now. We have managed to revive the most important of the Lorentz transformations – this time not only by half, but the transformation as a whole:

$$t' = \left[t - \frac{vx}{c^2}\right]\gamma. \qquad (2.39)$$

Now t', in our eyes, is not just a fictitious variable, helping to solve Maxwell's equations (though in that function it can also serve well), but the time, measured by the moving clocks. If those clocks are spatially separated, they must be synchronized through some real signals (for example through rays of light) or by means of a slow transportation from one point of space to another.

The physical essence of the transformation (2.39) can be reduced to the following two statements that may be regarded separately though they depend on each other:

1. **If a clock is moving with a speed v, its tick is retarded in inverse proportion to γ.**
2. **If two synchronized clocks, moving with the same speed v, are separated by a distance x along the direction of motion, then the clock placed first will be slow, as compared with the clock placed next, by the value $(vx\gamma)/c^2$.**

To draw the first conclusion from formula (2.39), it is enough to make there a substitution $x = vt$. It's just the way how the values x and t are connected with each other in the case of uniform velocity. Thus, we obtain $t' = t/\gamma$, which means t

hat the tick of a moving clock is slowed in comparison with a stationary clock.

The second conclusion can also be deduced from the formula (2.39), after a fixed (constant) value of t (for example, $t = 0$) is substituted there. In doing so, we want to compare the readings of different clocks that are located at different places with the different values of x at the same moment of time t. In other words, we want to have a snapshot of many moving clocks.[1] Having substituted $t = 0$ into (2.39), we arrive at $t' = -vx\gamma/c^2$. These two conclusions can be converted into each other by means of the slow transportation of moving clocks. In other words, taking conclusion 1 as a starting point and using the slow transportation of moving clocks, we can come to conclusion 2.

Now that we see how the moving rods and clocks behave, we can use that knowledge to discover a lot of interesting information. Some of it will be given in the next two sections.

---

[1] Of course, this snapshot cannot be made with a single wide-angle camera located at some remote position aside of the trajectory of the moving clocks, because the difference in times it takes the optical images of different clocks to reach the camera, will inevitably distort the picture. To avoid such a confusion, the "snapshot" involved is to be synthesized from many local snapshots made simultaneously by different cameras, each of them located in the vicinity of the relevant moving clock.

2.5. Moving against the ray of light

where, being fixed to the ether, we will watch how the observer, moving through the ether with a speed v, against the ray of light, propagating with speed c, will measure his or her speed with respect to the ray and arrive at a surprising result: his or her speed with respect to the ray will prove c and not c+v as might be expected from the first sight. It is the instruments moving together with the observer and distorted by motion through the ether that are responsible for that curious result

2.5.1. When does the law of relating velocities retain its classical form?

Imagine a ray of light, propagating through the ether from the left to the right with a speed c. Let there be an observer, moving against that ray with a speed v. What is the observer's speed relative to the ray?

Electrodynamics tells us that the ray of light is a train of crests and troughs that propagate in a single file with a speed c. Let us mark one of the crests and make it more exact what a speed we are looking for. The observer's speed u, relative to the wave, is nothing else but a change of the separation between the observer and the marked crest of the wave per unit of time. It is obvious that this speed is equal to the sum of two velocities – the velocity of the wave and that of the observer:

$$u = c + v. \qquad (2.40)$$

Don't look for any trap here. Everything goes on as usual. It is important to keep in mind that all three velocities u, c, and v are measured with the same set of instruments, which in this particular case are assumed to be at rest. As long as this assumption is valid, the law for relating velocities has a familiar form (2.40). If v is close to c, then u is close to $2c$.

Perhaps you are puzzled by the fact that the speed u exceeds c. How could it happen? No material body can outrun light, because the mass of the body grows unboundedly when its speed approaches the speed of light. But has anyone outrun light in our scenario? Light propagates with the speed c as it should; the observer moves with a speed v, smaller than c; and it is only the separation between the observer and the marked crest that contracts with the speed u higher than c. Two material bodies have the right to move against each other with their speeds approaching the speed of light, but not exceeding it. The separation between them is then reduced at a speed that is near to $2c$. If an observer is one of those two bodies, we wonder what he thinks about it. Doesn't he imagine that the body running against him moves faster than light?

To consider something in the way physicists are usually doing it, the moving observer must have some data at his disposal (for example the results of some experiments or measurements) which could serve him as a starting point for his own reasoning. If there are no such data, his reasoning would be pointless. Suppose he has the same data at his disposal as we have, i.e. the speeds c and v, measured by our instruments which are at rest (with respect to the ether[1]). Assuming these data as reliable, he would arrive at just the same conclusions as ours. He would write relation (2.40) and would think of himself just the same as we think of him. When a second body is moving against him, he cannot suspect that body of outrunning light, because he has not measured the speed of that body by himself, and has only the data obtained by us. According to those data, he himself, as well as the second body, are moving more slowly than light. In other words, his motion cannot serve him as a reason for inventing anything new. Though motion causes the Lorentz contraction of

---

[1] If we ignore the ether, nothing will change in our forthcoming consideration given in this section. That's why we mention the ether in parenthesis. In the next section, we will be able to say farewell to it.

2.5. MOVING AGAINST THE RAY OF LIGHT

length and time in his brain, this may (and does) affect only the quickness of his mental activities, but not their eventual result.

2.5.2. How a moving observer can measure a one-way speed of light

For the moving observer to take the floor and report from a point of view different from ours, he must change the starting point of his speculations, i.e. he must change his set of measuring instruments. He can use for example his own measuring instruments, whose readings may be different from those of our stationary instruments. Such a step on his part would be quite reasonable – especially if he or she doesn't know anything about the ether and regards his or her instruments as being at rest, and our instruments as being in motion. If his own instruments showed that the body, flying against him, had a speed greater than c (not with respect to any other body, but with respect to those instruments), then that body would have in fact outrun the light. And if, measuring the speed of the marked crest, while flying against him, he managed to get the value u, determined by the expression (2.40), he would have then the right to announce to everyone that a great discovery was made: an electromagnetic signal has been found with a speed of propagation exceeding the speed of light.

So, let our observer start his activities. To make his actions more spectacular, we will place him on a long platform that will move together with him with the speed v against the ray of light. Meanwhile we will keep reporting his actions, while watching him by means of our instruments that are fixed to the ether.

The marked crest is quickly approaching the observer. To measure the speed of that approach, the observer will mount two posts on his platform: the front post A and the rear post B, a distance l apart along the direction of motion. While measuring this distance, the observer will make a "mistake", and will regard the distance as γl, because the length standard used by him is by a factor of γ shorter than our stationary standard. After that he will take two timepieces A' and B' and trigger them simultaneously at the post B. On leaving timepiece B' at post B he will slowly and carefully transport the second timepiece to post A. The preceding section tells us that this operation will make timepiece A' lose time with respect to timepiece

B' by the value $\beta l\gamma/c$. During the transportation, its net speed of motion is a bit higher than v and its hands are therefore moving a bit more slowly than the hands of the clock B'. This bit is quite enough to cause this delay, which the observer himself does not notice. He may even place a TV receiver halfway between the two posts and, glancing at the images of the two timepieces on the screen, make sure that they show the same time. The shift of the hands of the clock A' in the process of its transportation remains unnoticed not only by the observer himself, but also by his instruments.

Both timepieces can be equipped with special automatics that will immediately stop each clock as soon as the marked crest of the light wave comes alongside the timepiece. With such automatics, the observer is free to go, say, to a library and enjoy there reading books on relativity. It should be so interesting to see what has been published about such experiments[1].

While the observer is in the library, the marked crest of the light wave comes alongside clock A'. The automatics works and the clock hand stops at some point t_A. At the same moment of time, according to our instruments, the hand of B' is at the point

$$t_B = t_A + \frac{\beta l\gamma}{c}, \qquad (2.41)$$

because B' is fast with respect to A' by $\beta l\gamma/c$ as was established above. Clock B' keeps on going, while the marked crest of the light wave keeps on flying from A to B. According to our instruments it is flying with the speed c, while post B is speeding against it with the speed v. The time $l/(c+v)$ is needed to the marked crest in order to reach post B. By this time the hands of clock B' will have moved $l/[\gamma(c+v)]$ points further, both the timepieces ticking by a factor of γ more slowly than our stationary clocks. As soon as the crest of the wave arrives at clock B', its hand stops to show the time

---

1 There is a lot of literature where such experiments are regarded as generally impossible. The ray of light does not return there to its starting point (so-called "one-way" experiments). The authors underestimate a possibility of matching the clocks by means of a slow transportation, which was proposed in a few decades after the development of special relativity and remained unnoticed in most textbooks.

2.5. MOVING AGAINST THE RAY OF LIGHT

$$t_B = t_A + \frac{\beta l \gamma}{c} + \frac{l}{\gamma(c+v)}. \tag{2.42}$$

The observer, who has returned from the library, sees the stopped timepieces show the times t_A and t_B and uses them to calculate the speed u' of the wave crest propagation as the ratio of the distance between the posts γl, measured by the observer earlier, to the difference between the readings of the two timepieces:

$$u' = \frac{\gamma l}{t_B - t_A} = \frac{\gamma l}{\dfrac{\beta l \gamma}{c} + \dfrac{l}{\gamma[c+v]}} = \frac{\gamma c}{\beta \gamma + \dfrac{1}{\gamma[1+\beta]}} =$$
$$= \frac{c}{\beta + \dfrac{1}{\gamma^2[1+\beta]}} = \frac{c}{\beta + \dfrac{1-\beta^2}{1+\beta}} = \frac{c}{\beta + [1-\beta]} = c. \tag{2.43}$$

So, the speed u' with which the observer and the crest of the light wave are approaching each other, proves to be equal just to the speed of light c rather than to the value $c + v$, obtained by the instruments fixed to the ether. Thus, the speed of light relative to the instruments by which it was measured, always proves to be equal to c, irrespective of the fact whether the instruments are at rest or in motion with a constant velocity. If the set of instruments moved not against the ray of light, but in the same direction as light, or perpendicular to it, or else at any angle to it, the result would be always the same $u' = c$. This rule is formulated in the following way: *The speed of light does not depend on the motion of the observer.* The word "observer" signifies here, of course, the set of instruments which are used to measure the speed of light, and not the man himself who, as it has been demonstrated, may be either at the library or somewhere else when the measurement was under way.

Independence of the speed of light from the velocity of the observer is part of a more general principle of the constancy of the speed of light, which also includes independence of the speed of light from the velocity of the source. The idea of absolute independence of light wave propagation from the source that has emitted it was born together with electrodynamics. At that time, light was regarded as traveling in the ether like the sound traveling through the air. Therefore the speed of light was supposed to be determined exclusively by the properties of the ether. But if the measuring instruments are moving through the ether against the ray of light, the speed of light, measured by them, seemed to be the sum of $c+v$. After the development of special relativity, it became clear that it was not so. The motion with a uniform velocity affects the length of rods, the tick of clocks and the difference in the readings of spatially separated clocks. These effects take place in a special way making the speed of light always equal to c. It might seem that it is the ether that is the cause of the changes in the properties of instruments moving through it. But for that explanation to be sound, it is necessary to have some proof of the existence of the ether. But once again, the clever ether covers up the traces, depriving us of every evidence of its existence. Should the ether really exist, the instruments fixed to it would give their own result of measurement which would be different from that registered by instruments moving through it. But nothing of the kind occurs! Whatever velocity the instruments travel with, the result of measuring the speed of light is always the same $u' = c$.

2.5.3. A general form of the law of relating velocities

Let us see now how the velocities will be related when it is not a ray of light, but a certain body C that moves against the observer. If the observer is moving with a speed v and the body, flying against him, is moving with a speed w, then the speed u of their becoming nearer to each other depends on the fact whether the measuring instruments are in motion or at rest. If all the three velocities: v, w, and u have been measured by the instruments at rest, they obey the ordinary law for relating velocities:

$$u = v + w \ . \tag{2.44}$$

that does not need any comments. The speed u may even exceed the speed of light. That should not puzzle us because neither the observer nor the body C make any attempt to outrun light. It would be quite different if such a speed was shown by the instruments that move

2.5. MOVING AGAINST THE RAY OF LIGHT

together with the observer at the velocity v. That would mean that the body C has the speed u' greater than c with respect to the instruments that are measuring the distance to the body at different moments of time. In reality, the speed u' is always less than c, and we will make sure of it right now. The procedure of measuring the speed u' with which the two bodies approach each other will be absolutely the same as in the case of a ray of light. Thus we will not repeat the description of it. The only difference is that in (2.42) and (2.43) the speed of the mutual approaching $c + v$ will change for $w + v$, after which the derivation (2.43) will take the following form:

$$u' = \frac{\gamma l}{t_B - t_A} = \frac{\gamma l}{\dfrac{\beta l \gamma}{c} + \dfrac{l}{\gamma[w+v]}} = \frac{1}{\dfrac{\beta}{c} + \dfrac{l}{\gamma^2[w+v]}} =$$

$$= \frac{1}{\dfrac{v}{c^2} + \dfrac{1 - v^2/c^2}{w+v}} = \frac{w+v}{\dfrac{wv}{c^2} + \dfrac{v^2}{c^2} + 1 - \dfrac{v^2}{c^2}} = \frac{v+w}{1 + \dfrac{vw}{c^2}}. \tag{2.45}$$

This result tells us that u' will never exceed the speed of light c for any values of v and w less than c. In the limit when both speeds of motion v and w approach the speed of light $v = w = c$, the speed u', according to this formula, will also be equal to c. It even remains to be c when only one of the two speeds v or w reaches the speed of light. Formula (2.43) was derived just for that particular situation.

Expression (2.45) is called the relativistic law for relating velocities. For the speeds that are much smaller than the speed of light, the item vw/c^2 in the denominator may be neglected, which brings us to the usual law for relating velocities (2.44). In the case of small speeds both sets of instruments – that at rest and that in motion – give almost the same values of u and u'.

When interpreting the relativistic law for relating velocities or making use of it, we have to bear in mind that the velocities, taking part in this law, have been measured by different sets of instruments. The added velocities v and w have been measured by instruments at rest, while the resultant velocity u' has been obtained by means of instruments in motion. That's why the result (2.45) looks so odd. But what is the motive for such a "misleading" representation? Why should we measure some velocities with instruments at rest, other ones with instruments in motion, and then establish a mathematical connection between them? This connection is sure to be far from simple. The cognitive significance of (2.45) is the following.

We are sure that no material body can outrun light. But unwillingly, one can be enticed with the idea of outwitting nature and getting a speed greater than that of light by accelerating not the body but the observer himself against the moving body. The formula (2.45) shows that this trick doesn't work. The speed of a body, if measured relative to the instruments being used in this measurement never exceeds the speed of light c. If we try to accelerate the body, then in the vicinity of the light barrier, its mass approaches infinity, but if the measuring instruments move against the body, then the readings of these instruments will change in such a way that the speed of the body relative to these instruments will be always less than the speed of light.

2.5.4. A revival of one more Lorentz transformation

Let us look at the formula (2.45) once again and compare it with the first of the three Lorentz transformations given under number (1.20) (See page 48). With a mere change in notation, they do not differ from each other. If instead of w we write down $-u_x$, these two formulas will differ only in the sign, which is explained by a difference in defining the directions in which the velocities u and v are assumed to be positive. It turns out that once again one more Lorentz transformation has been "revived" by our considerations. The velocity u' in the transformation (1.20) has proved to be not a fictitious mathematical quantity without any physical meaning, but the velocity of a body measured by a set of instruments, that are moving in the same direction as the body, but with their own velocity v, different from the velocity of the body w. The other two transformations

2.5. MOVING AGAINST THE RAY OF LIGHT

(1.20) refer to the case with the velocities v and w perpendicular to each other. These transformations can be "revived" similarly.

What has been related can be summed up as follows:

1. **The speed of light does not depend on the motion of the instruments, used to measure its value, and is always equal to 300,000 km/s.**
2. **If the instruments which measure the velocity of a body are set in motion against the body, then the velocity measured will increase, but its value will never exceed the speed of light.**

There is just one more step left that separates us from Einstein's postulates. We have seen how motion affects the properties of rods and clocks in the eyes of rods and clocks that are at rest. But we do not know how stationary rods and clocks look like in the eyes of moving instruments. The answer to this question will be given in the next section.

2.6. Which meter stick is more reliable - that at rest or that in motion?

where the ether fades away and we at last are left face-to-face with inevitability of Einstein's postulates

2.6.1. Planning a search for the ether

In the previous sections, we made sure of a lot of miraculous conversions that happen to material bodies when they are set *in motion with a uniform velocity:* their *length contracts in the direction of motion, the tempo of the processes slows down, the time zones are formed* changing each other along the motion, *the transverse forces relax, the mass increases,* etc. All these changes can be observed and registered by instruments at rest. But once these instruments are set in motion together with their objects of measurement, their properties abruptly change and they fail to notice even a slightest trace of those wonderful conversions. It could be supposed, of course, that moving instruments have a very good reason for missing these wonders. They are corrupted by motion, which makes their readings wrong. As for stationary instruments, they are not affected by motion and can see all the wonders of moving worlds "in their true light". It seems that we should trust the instruments at rest, and not the instruments in motion. But if it were true, the instruments at rest would then be unique. There would be an infinite number of sets of instruments, that would have various constant velocities, and all of them would be "wrong" with the exception of only one of them which would be at absolute rest and would be thus giving the true results. In other words, among all the standard rods which are in motion with different uniform velocities, there would be only one –

2.6. WHICH METER STICK IS MORE RELIABLE – THAT AT REST OR THAT IN MOTION?

the longest of all – which would move with a zero speed and would be at absolute rest. The same would be then applicable to clocks. Among all the standard clocks which are moving with different constant velocities, there would be only one clock, that would have the fastest tick and could be therefore regarded as being at absolute rest. But if the absolute rest indeed existed, it would be then quite natural to associate it with the world ether, so that the uniqueness of the stationary rods and clocks could be explained by their being fixed to the ether. The velocity of any body should be then measured with respect to the ether without paying any attention to other moving bodies. Because this velocity would not depend on the motion of other bodies, the motion with uniform velocity, as well as the state of rest, would become not relative, but absolute.

Had the ether and absolute rest existed, they could be detected experimentally. To do that, it would be enough to compare all the moving standard rods and find out the longest. Then the longest rod could be declared to be at absolute rest, while all the other ones would be regarded as being in absolute motion with the relevant absolute velocities ascribed to each of them. Among all inertial frames of reference there would be a privileged one which would be fixed to the ether. In all other frames an "ether wind" would be blowing, whose existence could be established and the velocity – measured.

Detection of the ether could be performed also by means of standard clocks. Having compared the tick of the clocks moving with different velocities, it would be possible to find out the clock with the fastest tick and declare it to be fixed to the ether. There would be also the third way of detecting the ether. If the ether existed, there would be no time zones in it. The transportation of a clock from one point of the ether to another would not cause its being slow or fast, provided this transportation was being made slowly and carefully. Comparing the readings of the synchronized spatially separated clocks in one frame of reference with the readings of the similar clocks in another frame of reference, it would be possible eventually to find out the frame that would have no time zones. That frame could be declared to be at absolute rest.

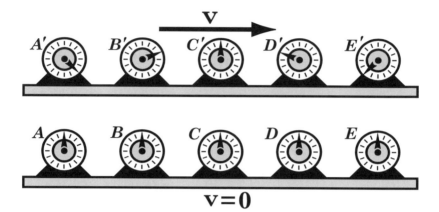

Fig.19 A family of synchronized clocks A', B', C', D', E' is moving from the left to the right with a uniform velocity **v**. Their readings are shifted in accordance with the time zones of their location in contrast with another synchronized family A, B, C, D, E which is at rest and is used as a reference for the estimations of the behavior of the moving clocks. If, however, we use the moving clocks as a reference for the estimation of the behavior of the stationary clocks, the readings of the latter will prove shifted in exactly the same way as the moving clocks are with respect to the clocks at rest.

2.6.2. The first failure

Let us try to do it mentally. We will begin with the third method because it is the simplest. Let us imagine that there is an ether, and all the spatially separated clocks A, B, C, D, and E, shown in Fig.19, are fixed to it. These standard clocks are believed to be at absolute rest. They have been synchronized with each other by means of a slow transportation, or, which is the same, by wireless time signals, whose delays have been taken into account. Suppose there is another set of similar clocks A', B', C', D', and E' moving to the right with a constant speed v. These clocks have also been synchronized with each other, say, by a slow transportation. Their hands have been set in the same position at the middle point C', after which the clocks have been slowly transported to their destinations. Section 2.4 tells us that the clocks, being transported forward along the motion, lose time in comparison with the clock left at the point C'. As for the clocks that are being transported against the motion, they, on the contrary, are getting fast. Let us assume that, at a cer-

2.6. WHICH METER STICK IS MORE RELIABLE - THAT AT REST OR THAT IN MOTION?

tain moment of time that corresponds with the disposition of the clocks shown in Fig.19, the middle clocks C and C' show the same time. Then the clocks D' and E', located ahead (with respect to motion), will be slow in comparison with D and E, while the clocks A' and B', on the contrary, will be fast as compared with their stationary counterparts A and B. Such a behavior of the primed clocks could be ascribed to their motion through the ether.

Let us turn now to the observer in motion. What will he see, looking at the readings of the clocks at rest and using his own clocks as standards? He will see the hands of the clocks C and C' in the same position. As for the clocks D and E, which are moving after C (he regards himself as being at rest, and the clocks at rest – as moving from the right to the left), they are fast in comparison with D' and E', while the clocks A and B are slow in comparison with their counterparts A' and B'. The moving observer will be sure to conclude that there are time zones in the system, which we have regarded as fixed to the ether. In other words, the moving clocks "think" about the stationary clocks the same as the stationary clocks "think" about the moving clocks. Absolute symmetry reigning here, it is quite impossible to say, on the account of the time zones, which set of clocks is fixed to the ether and which is moving through it. The observer "at rest" and the observer "in motion" can argue a lot about who of them is in motion and who is at rest. Comparing the readings of the spatially separated clocks "at rest" with the readings of the similar clocks "in motion", it is impossible to establish which of the observers is right.

2.6.3. The second failure

Let us now switch over to the tick of the clocks. Perhaps it is there that some asymmetry will be found between the clocks in motion and at rest. It seems that it is only the moving clock that is ticking more slowly when moving through the ether. Let us take two synchronized clocks A and B fixed to the ether, a distance x apart, as shown in Fig.20(a). Suppose that there is also a third clock A' that moves through the ether from the left to the right with a constant speed v. When A' comes alongside A, we make its hands matched with the hands of A. For simplicity, let both of them show a zero time at that moment. Some time later A' will come alongside B. At that moment B' will show the time x/v, while A' will read the time $x/(\gamma v)$, because it moves through the ether and hence its tick is slower by a factor of γ. Having compared the reading of A' with that of B, any observer, being either at rest or in motion (or even a foreigner), will then see that A' is slow, in comparison with B, by the value

$$\Delta t = \frac{x}{v} - \frac{x}{\gamma v} = \frac{x}{v}\left[1 - \frac{1}{\gamma}\right] = \frac{x}{v}\left[1 - \sqrt{1 - \beta^2}\right]. \quad (2.46)$$

The value (2.46), if measured experimentally, testifies that the clock in motion ticks more slowly than the clock at rest. In order to obtain this result we had to make use of two spatially separated standard

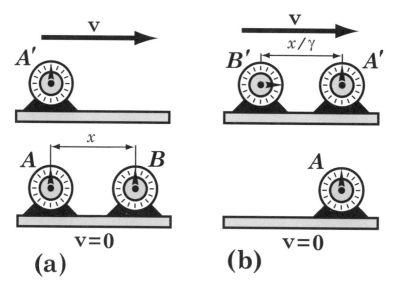

Fig.20 (a) To measure the slowdown of the tick of the <u>moving</u> clock A', we need two synchronized spatially separated <u>stationary</u> clocks A and B in order to make sure that A' shows the same time as A when they are alongside each other, and to register the difference in readings between A' and B as soon as A' comes up with B. (b) To measure the rate of the <u>stationary</u> clock A, we need two synchronized spatially separated <u>moving</u> clocks A' and B' in order to make sure that A shows the same time as A' when they are alongside each other, and to register the difference in readings between A and B' as soon as A comes alongside B'. Because A' and B' are in different time zones, it turns out that, in spite of the slowdown of their tick, clock A appears to them as ticking more slowly than any of them – the result which mirrors the case (a).

2.6. WHICH METER STICK IS MORE RELIABLE - THAT AT REST OR THAT IN MOTION?

clocks that were at rest (the clocks A and B), and one clock that was in motion (the clock A').

Let us see now what will happen when an observer, moving through the ether, repeats just the same measurements in order to learn how a clock at rest behaves in the eyes of the clocks in motion. We will report his actions and comment on them from the ether. The moving observer will have to provide himself with two standard clocks A' and B', a distance x apart, as shown in Fig.20 (b). The distance x will be measured by him in the same way as it has been done by the observer at rest. In other words, he will establish the distance x with his own tape-measure. The tape-measure, being shorter by a factor of γ, he will be "mistaken", and instead of x he will set the spacing x/γ between the two clocks. Placing the clocks A' and B' the distance x/γ apart, he will be sure that the space between them is x. On synchronizing the clocks A' and B', the moving observer will regard their readings as identical, while "in fact" they will prove to be in different time zones, and therefore B' will be fast compared with A' by the value

$$\Delta t_1 = \frac{\beta \gamma x}{c} \frac{1}{\gamma} = \frac{\beta x}{c} \qquad (2.47)$$

as determined by expression (2.32). (See page 132.) By that very value B' would get fast as compared with A', if it was slowly transported the distance x/γ against the motion through the ether, after it had been matched with A'. When A' comes alongside A, the latter is matched to A', so that their hands coincide. For simplicity, we will assign a zero value to these initial readings. Some time later B' will come alongside A, which will show the time $x/(\gamma v)$. By that moment, the hand of B' will have advanced by the value Δt_2, which is by a factor of γ smaller:

$$\Delta t_2 = \frac{x}{\gamma^2 v} ,$$

because the tick of B' is γ times slower. When A' was alongside A, the clock B' showed the time Δt_1 determined by the formula (2.47). Now the hand of this clock has moved Δt_2 points further. Thus, at the moment when B' and A are alongside each other, the hands of A

will read $x/(\gamma v)$ points, while the hands of B' will show $\Delta t_1 + \Delta t_2$ points. This means that at the moment of meeting between B' and A, the clock A will be slow with respect to B' by the value

$$\Delta t = \Delta t_1 + \Delta t_2 - \frac{x}{\gamma v} = \frac{\beta x}{c} + \frac{x}{\gamma^2 v} - \frac{x}{\gamma v} =$$

$$= \frac{x}{v}\left[\beta^2 + \frac{1}{\gamma^2} - \frac{1}{\gamma}\right] = \frac{x}{v}\left[\beta^2 + (1-\beta^2) - \frac{1}{\gamma}\right] = \quad (2.48)$$

$$= \frac{x}{v}\left[1 - \sqrt{1-\beta^2}\right].$$

Having compared this result with the formula (2.46), we see that according to the evidence obtained by the moving instruments, the stationary clock A is slow as compared with the moving clocks by exactly the same value by which the moving clock is slow as compared with the stationary clocks.

In spite of the presence of the ether, everything proved to be quite symmetrical. The thing is that the clocks A and A' are unable to compare the rate of their ticking without making use of at least one additional clock. This is because A and A' turn out to be at one and the same place only once and never meet again either in the future or in the past. Thus the result of comparing the rates of the clocks A and A' depends on what clock will be chosen as the third participant of the experiment. If it is the stationary clock B, then the tick of A' will prove to be γ times slower than that of A. But if it is the moving clock B' that is taken as a third participant, then everything will turn over symmetrically and, this time, the tick of A will prove slower. In other words, before comparing the rates of the clocks A and A', one of them is declared to be a standard, and the other – an object of measurement. Because one standard clock is not enough to make the measurement, the standard clock must be completed with one more auxiliary standard clock, placed a certain distance apart from either A or A'.

So, among many clocks moving with different velocities, there has been none that would have the fastest tick. The measuring clock, completed with an additional standard clock, will always be faster than the clock measured irrespective of the fact whether the clock being measured is moving through the ether or is fixed to it. The ether proved to be invisible once again.

2.6.4. The last failure

So the clocks, whatever combinations they might be used in, failed to detect the ether. Can rods be capable of doing it? If the ether does exist, then among numerous rods of standard length, moving with different velocities, there must be one that would be the longest. The very fact of the existence of such a rod could be a good proof of the presence of the ether. But in order to find the longest rod, it is necessary to develop a method of comparing the lengths of two rods, moving relative to each other. To compare the lengths of two rods, moving relative to each other, it is not sufficient to have the two rods, it is also necessary to have spatially separated clocks. When the zero marks of the two rods coincide, just for one short moment, it is necessary to see where at that very moment the ends of those rods are. But the thing is that the words "at that very moment" do not denote the same for the stationary observer and for the observer in motion. Each of them has his own system of spatially separated clocks, and what is simultaneous for one of them is not simultaneous for the other.

Again it's those spatially separated clocks that mix everything up! They are a stumbling block to all our maneuvers. Once they appear on the boards, a flood of wonders starts pouring out as if from the horn of plenty. Can it happen now, for all that we know, that the moving length standard A' is shorter than the standard A, which is at rest, and at the same time the standard A is shorter than the standard A'? There is no point in concealing from you that it will be just the result awaiting us here – at the end of our rather long investigation. In the eyes of an observer fixed to the ether, this paradoxical result might be commented on in the following way. One of the two standard rods involved is used as a measuring instrument while the other is an object of measurement. There arise two situations which seemingly contradict each other. In the first case, the rod A is used as a measuring instrument. In the second case the functions of the rods are interchanged – it is A' that is used as an instrument while A is just an object of observation. But it is not a wording that is of the main importance here. Both rods are equipped in accordance with their functions. The measuring rod (either A or A') is always equipped with two spatially separated clocks TT or $T'T'$ (it would be impossible to compare the two lengths otherwise) while the object of the measurement does not need any additional equipment (that rod is nothing else but just an object to be measured). When A' is measured with A (the first case), it is done with the set of instruments $A + TT$, while in the second case (when A is measured with A') it is done with the set $A' + T'T'$. Remember that the instruments $A + TT$ are at rest (and not subject to any deformation), while $A' + T'T'$ are in motion (and inevitably distorted by that motion). The two different sets of instruments (in motion and at rest) with different properties are used in those two cases. It is a good reason for arriving at a paradoxical result, isn't it? It is the spatially separated clocks (two clocks at least) that are responsible for everything. They are at rest in the first case (showing the same time at different places) and in motion in the second (showing different times at different places), and we have no means to avoid this "discrepancy". As a result of that "discrepancy", the rod equipped with the clocks is sure to be longer irrespective of its state – whether it is in motion or at rest, provided the moving rod is measured with the instruments at rest while the rod at rest is measured with the instruments in motion. We could speak at length about it, giving every detail, but perhaps it would be too bothersome. So we will confine ourselves to some general remarks.

So, according to stationary instruments, the moving rod is shorter than the stationary one, while according to moving instruments, the stationary rod is shorter than its moving counterpart. On the face of it, it seems paradoxical. But we come across such "paradoxes" in our life much more often than we might have believed. Suppose you are walking along the street and run into a friend of yours. After meeting, greeting and by-passing each other you stop and begin talking. First you exchange impressions of the circumstances of your meeting. "You have passed by me on my left", – your friend says. "Oh no, – you say – it is you who has passed on my left." In this situation both of you are right. Just the idea of "the right" and "the left" depends on the direction the man is looking, while saying those words. It is enough to turn 180^0 round and the words will get the opposite meaning.

A non-trivial situation we have been entrapped in while studying the effect of motion upon the properties of bodies has at last become understandable. There is no rod in nature that could be declared the longest! If someone tells you that he has managed to

2.6. Which meter stick is more reliable - that at rest or that in motion?

find the standard rod which is the longest among all the other moving standard rods (in other words, that he has managed to detect the ether), you will say to him at once: "Let us have this longest rod! We will shorten it immediately, without even touching it. We will just take an arbitrary set of measuring instruments in motion and will measure your rod with them. Looking at the results of our measuring, you will have chance to make sure that your rod is shorter than ours."

2.6.5. Summary

Summing up what has been said here, we arrive at the following rule:

All the moving sets of measuring instruments are equal in their rights. None of them may be declared to be at absolute rest or in absolute motion

It is only now that we are in position to assimilate Einstein's famous postulates quite consciously. The next section is devoted to them. Being equipped with those postulates, we will again return to our rods and clocks and demonstrate their behavior in some special situations.

2.7. Einstein's postulates

where Einstein's postulates are given together with some tips about their usage

2.7.1. The postulates

More than a century ago, after the main laws of electrodynamics had been successfully formulated, everyone believed that Galileo's principle of relativity lost its universality. The propagation of light and other electromagnetic phenomena did not get along with that principle and it seemed that they could not take place without a certain medium, the universal and all-pervading ether which is at absolute rest and relative to which the uniform velocity of any motion should be counted. The greater this velocity, the greater the influence of the ether on processes going on in the moving system. Because the influence of motion with uniform velocity on various phenomena in a moving system did not raise any doubts, detectibility of the ether was out of discussion, and in spite of the first failures everyone was sure that sooner or later the "ether drift" would be detected and measured experimentally. Everyone knew that, even at uniform velocity, the motion affects processes in the moving system in a rather odd way. To make sure of it, it was enough to glance at the Lorentz transformations. But the reality exceeded every expectation. This influence proved to be so sophisticated that it excluded any chance of detecting even a slightest puff of the "ether wind". We could make sure of it in the previous sections. No matter how peculiar the moving systems or objects behaved, the result was always the same: the ether as well as absolute rest proved to be invisible ghosts that did not lend themselves to experimental observations. Beyond the odd behavior of equilibrium systems one could feel the almighty hand of nature that subjugated everything to the universal principle of relativity of motion.

In the previous sections, we have passed a long way to approach this principle, investigating the detailed mechanisms of the phenomena caused by motion. Our investigation was based exclusively on the well-known laws of electrodynamics and mechanics as they had

2.7. EINSTEIN'S POSTULATES

been formulated by Newton, Maxwell and Lorentz. We did not resort to any additional suppositions or postulates. Eventually we got convinced that there is no absolute motion in nature, and that it is impossible to detect the ether. Einstein had built his Relativity in the inverse order. He had figured out everything at once. He just postulated the relativity principle, which we managed to reach through a lot of hard work. More exactly, he formulated two famous postulates which, in a somewhat simplified form, can be worded in the following way:

1. ***Changes in the states of physical systems obey the same laws in all inertial frames of reference.***
2. ***The velocity of light does not depend on the motion of the source.***

Let us first consider the physical meaning of the second postulate. It agrees with the pre-Einsteinian ideas of Maxwell-Lorentz', according to which light propagates through the ether, and it is therefore quite natural that the velocity of light does not depend on the motion of the source.[1] But it could depend on the motion of the observer through the ether if it were not for the first postulate that excludes this possibility. Apparently, the two postulates contradict each other. That's why prior to Einstein no one even dared to think of uniting them in one theory. Now you are aware of the gist of it. The moving rods and clocks behave in such a way as if they conspired to reconcile these two postulates. Before Einstein, no one had suspected it. As for Einstein, he deduced the properties of the moving rods and clocks not from the well-known laws of nature, but from his own postulates. In doing so, he demonstrated that they do not contradict each other. He has just required of rods and clocks to obey his postulates. And lo and behold, the rods and clocks not only have obeyed his demands, but they have done it in full agreement with the well-known laws of nature.

From a formal point of view, Einstein acted like this. First he required of the velocity of light to be independent of the motion of

---

1. In most textbooks on relativity, this postulate is presented in an oversimplified form — having been replaced by the law of constancy of the speed of light which, in addition to Einstein's endeavor, includes the independence of the speed of light from the motion of the observer. The reader may find many interesting details about it in the historical review at the end of this book. (See pages 240-243.)

the observer. Simple mathematical derivations led him to the Lorentz transformations (1.11)–(1.13) for length and time. (See page 48.) In these transformations, the variables x' and t' (especially t') acquired a new physical meaning that Lorentz had not ascribed to them. They were interpreted as the results of the measurements of length and time, made by the moving instruments. It turned out that the transfer from rest to motion makes rods and clocks contract their length and slow down their tick, so that the time zones appear in any system of moving bodies. This result was the cornerstone of relativity and was named by Einstein "Relativity of Simultaneity of Spatially Separated Events". Einstein did not go deep into the particular reasons of such an "odd" behavior of rods and clocks. He was so sure of the validity of his initial postulates that he just did not see any necessity to do it. He believed that every rod and every clock are sure to have their own good reasons for the length contraction or time slowdown. Such approach was justified by the fact that Einstein had declared the most general principle that involved all the branches of natural science in the present, past, and future.

Having turned over all ideas of space and time, Einstein focused his attention on mechanics and demonstrated that the new ideas formally got along with the former laws there, provided the mass was assumed to depend on the velocity according to the formula $m = m_0\gamma$, from which it followed that energy and mass are equivalent to each other: $W = mc^2$.[1] It also turned out that forces must be transformed according to the rule (1.39). (See page 63.) Einstein then returned to electrodynamics, found a number of relativistic effects there (some of them will be considered in the next section) and gave new, relativistic interpretation to some phenomena known before. The limits of the book do not allow to give a full account of what had been done by Einstein in the process of developing special relativity. We will therefore confine ourselves to responding to some general questions and doubts that arise inevitably when Einstein's approach is considered for the first time.

Question 1: *Why did Einstein need the second postulate, whereas all the main results could be deduced from the first postulate?* Because the first postulate, when applied to particular physical situations, inevitably uses the speed of light c as a very important

---

1. This does not refer to Newton's law of gravitation which is out of the scope of special relativity.

2.7. EINSTEIN'S POSTULATES

coefficient. The first postulate guarantees its independence from the motion of the observer. But it does not guarantee the independence of c from the motion of the source. Let us suppose that c does depend on the motion of the source. It would then turn out that the relations, describing a fundamental connection between space and time contained a coefficient that depended on such an accidental thing as the motion of the source. Moreover, it would be even not clear what source was meant if we spoke, say, of the length contraction of a moving rod. It would be possible, though, to do without the second postulate, deducing it from the laws of electrodynamics, but Einstein wanted all the properties of space and time to follow only from his postulates extending far beyond the limits of electrodynamics. And the speed of light in relativity is not just the speed of the propagation of electromagnetic waves, but a world constant, that serves also as the upper limit of the speed of motion of any material body. That's why the second postulate was required, though it played an auxiliary role to the first postulate.

Question 2. *Is it possible to deduce another special relativity (different from that of Einstein's) from Einstein's postulates?* At least several relativistic theories, different from Einstein's approach, have been published in scientific literature. In most of them the properties of the world space are direction-dependent, though there is no evidence which might either prove or disprove such propositions. Unlike Einstein's special relativity, all those theories stand apart from the main laws of physics. The relativistic effects, predicted by them, require some corrections in the laws of mechanics or electrodynamics, which is perhaps the main reason why those theories are not widely known or readily recognized.

2.7.2. An event and its description

This *question is caused by the fact that one and the same event is seen differently by observers (instruments), moving with different constant velocities. How great can be this difference in the perception of the same events? Can it happen that according to one observer a certain event has occurred, while according to the other – it hasn't?* This can never happen, of course. If some event has occurred (a supernova has exploded, a lightning has struck a tree, a living being has been born or died, etc.), it has occurred for every observer, and sooner or later all the observers will be able to register it by their instruments or through their sense-organs.[1]

Nevertheless the coordinates of an event and the moment of time when it happened are not the same for different observers. There are two reasons for it. The first of them does not directly refer to relativity. Just every observer can have his own reference point from which space and time are counted. Such a kind of "relativity" had been known well before Einstein developed his theory. The second reason is associated with relativity: the length of the rod and the rate of the clock depend on the speed of their motion. It is not only the coordinates and time that are different. Many other physical quantities are different too – the forces of interaction and the masses of bodies, taking part in the event, their relative velocities and accelerations. Even the reasons explaining the event may be different. But the very fact that the event has occurred cannot be concealed from the instruments just because they are in motion or at rest.

To illustrate this idea let us look at the electric charge moving with a uniform velocity. A moving charge is a kind of a current. Therefore there is a magnetic field around the trajectory of the charge. If we take a stationary compass, its pointer will be deflected and will indicate the direction of the lines of the magnetic field. From what has been just said, it is only the deflection of the pointer that is an event. Everything else is just a description of the circumstances under which this event has occurred, and an explanation of the causes that are responsible for it. Now let us see how this event is interpreted by the observer who is moving together with the charge. In the view of this observer, the charge is at rest. There is an electric field around it, but there is no magnetic field. The compass is moving and its pointer certainly turns (the event occurs irrespective of the observers), though there is no magnetic field there. The pointer shows the direction of the magnetic lines that do not exist... Of course, this "explanation" doesn't work because it is true that for the observer who accompanies the charge there is no magnetic field at

---

1. There may be, though, one exception. It refers to the events which occur within so called "black holes" – supposed accumulations of matter in cosmos, so large and concentrated, that the force of gravity, arising there, does not allow even the rays of light to escape from there. The events which have occurred there will never become known to anyone who is outside the hole. A fascinating story about their origination and properties has been presented by Kip Thorne in his book "Black Holes and Time Warps", Norton & Company, 1994.

2.7. EINSTEIN'S POSTULATES

all. (We are sure of it because the instruments he has got with him will not be able to register any field.) What then makes the pointer turn? If the observer, equipped with special devices, peeped inside the pointer which moves relative to him with a constant velocity, he would be greatly amazed by curious things happening there under the action of the pointer's motion. A boundary is formed across the whole pointer from its north pole to the south pole, with electrons piled up on the one side of this boundary and deserted from the other side. (The net charge within the pointer is, certainly, conserved.) One half of the pointer proves to be charged negatively, the other half – positively. The stationary charge (that can create only an electric field) attracts one half of the pointer and repels the other half, which makes the pointer turn.

For the observer that is fixed to the compass this redistribution of the charge remains invisible. His instruments do not register it. If he even tried to count the electrons just by his finger, he would find them in equal amounts in both halves of the pointer. But the second observer (relative to whom the pointer is in motion), using the same method of measuring, i.e. counting the electrons by his finger, would find different numbers of electrons in the two halves of the pointer. In the next section we will examine everything in detail, so now we will simply mention that the electrons inside the pointer are always in motion, and their content in any half of the pointer is continuously renewed. Some electrons escape to the other half, and the same number of electrons return, so the net number of electrons within a given volume remains the same. Thus, either of the observers has to count not just the number of the electrons in one half of the pointer, but the number of the electrons that are in different points of the pointer *simultaneously*. But for our two observers, the idea of simultaneity is not the same. What is simultaneous for one of them is not simultaneous for the other. That's why they will count up different numbers of electrons in the same half of the same pointer.

You might be puzzled: isn't the redistribution of electrons an event? And if it is, then why can't it be noticed by one of the two observers? Your arguments would have been quite reasonable if the redistribution of electrons had arisen indeed. But it had not arisen since the beginning of our observations. It had been just present there. It had existed there both before and after the pointer turned. It had occurred long ago – as early as the compass was set in motion – and it had remained constant since then. According to any observer it will remain constant further on – until the uniform velocity of the compass suffers a change. It will be only then that the distribution will change, and this change will be an event, noticeable to any observer, either in motion or at rest. But as long as the pointer is moving with constant velocity, the distribution of charge is constant, so no events take place there.

Summing up, we may give the following definition: *An event is a result of an interaction of bodies or of other physical objects* (such as, say, an electromagnetic field). If there were no events at all, everything would move just by inertia, and even accidental collisions between the moving bodies would be impossible, because every collision is an interaction. Then, coming across each other, physical objects would have behaved like shadows or ghosts – they would not notice each other and would pass quite indifferently through each other. Such a world would be dull and boring. There would be no events in it. Physics as a science, would have vanished being deprived of both the objects and the means of investigation. None of the instruments would be able to measure or register anything, because any measurement is based on the interaction between the instrument and the object of the measurement and is, consequently, an event. Luckily, nothing of the kind happens in reality. The world is full of events, that are taking place even more often than we would like them to.

So, events are absolute (they are registered in all frames of reference), while the way of describing them is relative (it depends on the choice of the frame of reference, or, to be more exact, on the velocity of the set of instruments through which the description is made). There is an important comment to it. The way of describing is relative as long as we use such customary physical quantities as "length", "time", "force", "energy", "mass", "momentum" etc. All these quantities depend on the choice of the frame of reference. But we can define other quantities which will not depend on the frame of reference, being the same for all the observers. Such physical quantities are called *invariants*, i.e. undergoing no variations.

Suppose there are two events that have occurred at two different points A and B at different times. Let us regard them from different frames of reference, that are moving with different uniform speeds along the straight line AB. We will choose any two of them. The coordinates of the events and the moments of time they have occurred depend on the frame of reference. In one of these frames, they can be

2.7. EINSTEIN'S POSTULATES

x_1, t_1 and x_2, t_2. In another frame they may be different. We will designate them as x'_1, t'_1 and x'_2, t'_2. By means of the Lorentz transformations (1.11) and (1.13) (see page 48), it is easy to show that the following equality holds:

$$c^2(t_2-t_1)^2 - (x_2-x_1)^2 = c^2(t'_2-t'_1)^2 - (x'_2-x'_1)^2 . \quad (2.49)$$

You can make sure of it by yourself. Take the expressions for the relevant primed values from the Lorentz transformations (1.11) and (1.13) and substitute them into (2.49). Equation (2.49) will at once turn into an identity. This means that the value

$$(\Delta s)^2 = c^2(\Delta t)^2 - (\Delta x)^2 , \quad (2.50)$$

called an interval, is the same in all inertial frames of reference, and is therefore an invariant. Other invariants are also known, but we will not dwell on them. The values Δt and Δx depend on the choice of the frame of reference, but not the interval Δs. What is the physical meaning of the interval? What mysterious physical objects in nature does it correspond to? We can't give a satisfactory answer to these questions now. There is an impression that both Nature and Relativity try to demonstrate something very important but we are not in position to comprehend it as yet.

Imagine a rod, hovering indifferently in a state of weightlessness, in the middle of your room. Let us project this rod on the floor of the room and on the two adjacent walls. Having measured the three components Δx, Δy, and Δz, we will be able to calculate the length of the rod Δr by means of Pythagorean theorem:

$$(\Delta r)^2 = (\Delta x)^2 + (\Delta y)^2 + (\Delta z)^2 . \quad (2.51)$$

Now suppose our room has turned, but not the rod. All the three components Δx, Δy, and Δz have changed. The readings of the instruments, measuring these components, are quite different now. But the value Δr, as defined by (2.51), has remained the same. It is the length of the rod, hence it cannot depend on the turning of the room. The same is valid if the rod is rotating within a horizontal plane ($\Delta z = 0$):

$$(\Delta r)^2 = (\Delta x)^2 + (\Delta y)^2 . \qquad (2.52)$$

Now compare equations (2.50) and (2.52). They have a lot in common, don't they? The speed of light in (2.50) is a fundamental constant, that can always be turned into unity by choosing accordingly the units of length and time. The only formal difference between (2.50) and (2.52) is the following: in the right part of (2.50) two numbers, known to be positive, are subtracted while in (2.52) they are added. This does not prevent, however, the values Δs and Δr from being treated in mathematics similarly, in spite of the fact that the squared variable $(\Delta r)^2$ is always positive, while the value $(\Delta s)^2$ may sometimes be negative. The essential difference between the quantities Δt and Δx lies in something else. In the case of a rod we may, instead of running the components Δx, Δy, and Δz just measure the length of the rod itself, and we know how to do it. Not so for the interval. We are not able to measure it directly. Moreover, we have no idea of what it is we want to measure. The interval is perceived by us only through its components Δt and Δx, and it is only they that, nowadays, can be the objects of our direct measurements. And they depend on the choice of the frame of reference. So it turns out that in our experimental work as well as in engineering, we are able to deal only with shadows on the wall (i.e. with space and time) and we have no idea of what casts these shadows. (Sometimes, however, we can descry the interval itself. This happens when it is oriented parallel to the shadow it is casting. Observing the shadow, we see that in this particular case it coincides with the interval itself.) Nevertheless it does not prevent physicists and engineers from making use of the interval in both theory and engineering. By means of invariants it is often easier to arrive at the simplest solution of the problem, that otherwise would have taken a lot of effort and time. Invariants therefore stand high in the eyes of physicists and are used by them very widely, though their mathematical representation is not always so simple as in (2.50). Some competent physicists even think that Einstein's choice of the name for his theory was not the best. It may seem that according to relativity everything in the world is relative, that there is nothing absolute in the universe. Which is not true. Relativity does not abolish all absolute quantities. It only replaces some old absolute quantities by new ones. Time had been regarded as absolute before. Einstein disproved this prejudice. But there appeared an interval Δs that proved to be absolute. No matter that its physical meaning still remains a mystery. Perhaps some time will be needed to get to the point.

2.7.3. Relativistic effects in their dynamics

It's time, however, to return to our "shadows on the wall", i.e. to time, space, and other familiar concepts and see once again how these shadows will be changing when the room is turning while we are switching over from one frame of reference to another. We mean the Lorentz length contraction and clock slowdown as well as all the other relativistic effects. Can it be that all of them are just seeming? When we see an event, everything is clear and there are no problems about it. If it has taken place, no one doubts it. Not so for the relativistic effects. Take, for example, the slowdown of the tick of a moving clock. This slowdown cannot be qualified as an event, because it happened long ago, when the clock was set in motion (if it really was and hadn't been born in the moving frame), and has not suffered any change since then. For the observer who travels together with the clock, this slowdown does not exist at all, while all other observers take it differently, according to the velocities of their motion. Doesn't it mean that the observer traveling together with the clock is always right – there is no slowdown indeed and all other observers are strayed off by the wrong indications of their instruments? Putting forward such a question, we are about to suggest that everyone should share the point of view of the observer who accompanies the moving clock. As for all other observers, we are inclined to deprive them of their right to judge objectively about the phenomena that take place in other frames of reference which are in motion relative to them. Kindly allowing them, though, to register the events occurring in other moving frames, we forbid them to interpret those events and believe only the opinion of the observer who accompanies the clock. Acting like this, we put the frame following the clock into a privileged position, which contradicts Einstein's main idea that all the inertial frames of reference must have equal rights. Such an approach is wrong even from a purely formal point of view.

Now let us see what the slowdown of the moving clock really means for the observer at rest. Suppose there is an interesting TV show in a moving world. We want to see it from another world which is at rest. From the interstellar bill, we know the frequency of the transmitting station, and, tuning in our TV to it, we suddenly discover that the screen is blank. The generator of the radio signal is in motion, and all the processes in it are therefore going more slowly. The frequency of the signal proved much lower than we had expected. If we ignored this effect as "seeming", we would not have been able to see the TV show that interested us. It will be another story if we take into account the relativistic effect and turn the tuning handle of our TV, adjusting it to the frequency which is lower by a factor of γ. The picture will immediately appear. The fact that we have turned the tuning handle is certainly an event that can be registered by every observer in all the moving worlds. Having discussed this fact at the interstellar symposium, all of them would have admitted that relativistic effects are not seeming, even in the situation where all systems keep on moving with constant velocities.[1]

With the change of the velocity of motion, relativistic effects are displayed still more brightly. Though Einstein's postulates pertain only to bodies which are moving with uniform velocities, special relativity very often deals with situations in which either the bodies under consideration or the measuring instruments involved change the uniform velocities of their motion. Let us return to the example with the TV reception. When we switched the TV on, the screen was blank. Whose fault was it – generator's or receiver's? We think it is the generator's fault, for it is the motion of the generator that caused the change in the frequency of the signal. The residents of the moving world do not agree about it. This should be expected because they do not notice any slowdown in the processes which take place in their frame. If judging from the readings of their instruments, the broadcast goes on at the frequency, announced in the interstellar bill. They believe that they are at rest, while we are in motion. By means of their instruments they find certain "defects" in our method of receiving the broadcast, ascribing these "defects" to our motion. Due to these "defects" the frequency of the signal at the input of our receiver proves to be below the proper value.[2]

---

1. For our mental experiment with the transmitter and receiver to exactly proceed the way it was described above, an important condition ought to be satisfied. The line connecting the generator and the receiver must be perpendicular to the direction of motion. Otherwise, though the effect remains, everything becomes much more complicated. If the duration of the experiment is short enough, then we can neglect the change in the angle at which the transmitter is seen by a stationary observer.
2. The explanation of these "faults" is too cumbersome to dwell on it here. We will just note here that from the point of view of the moving observer the line, extending from the observer to the transmitter, is inclined to the perpendicular to the direction of motion. Our instruments do not notice this inclination.

2.7. Einstein's postulates

So which of us is right — they or us? In this situation both are right, because it is impossible to tell which of us moves and which is at rest. Both points of view are irreproachable. And though the arguments of the parties differ to a great extent, the conclusion is the same — the tuning handle of the receiver must be turned toward lower frequencies by a certain, quite definite value. Let us change the initial conditions in our mental experiment. Let the transmitter and receiver move parallel to each other with the same speed. It is possible to say that both of them are at rest. The reception is perfect — exact at the frequency, indicated in the bill. Let us suppose now that the transmitter increases its speed of motion. The frequency of the signal emitted by it, will decrease and the picture on the TV screen will vanish. Which of them is to be blamed now — the transmitter or the receiver? This time, the answer is quite definite — it is the transmitter. Relativity does not allow us to distinguish the state of the motion by inertia from the state of rest. Therefore the moving observer cannot say whether he is at rest or in motion, and if in motion, with what velocity he is moving. There is no absolute motion by inertia in nature. But the change in the velocity can be measured by the moving observer as exactly as he likes. He doesn't know the absolute velocity of his motion either before or after the acceleration. But the net change in the velocity can be registered by his instruments. He can use the forces of inertia that are exerted during the acceleration. Having measured them, he could find the accelerations as a basis for evaluating the change in the velocity.

2.7.4. *Elasticity from a relativistic standpoint*

When the velocity of a system changes, we must distinguish two essentially different cases. In the first case, there are no isolated parts within the system. Any particle (or a group of particles) inside the system occupies a certain, quite definite equilibrium position (or maybe oscillates or rotates about this position). In the state of equilibrium all the long-term forces exerted on a given particle (or a group of particles), cancel each other. But as soon as the particle tries to leave that position, there arise forces that return it back. If some particle or part of the system experiences short-term local oscillations or rotations, then the equilibrium of the whole system is possible only for a certain frequency of those oscillations. Should this frequency change for a certain reason, there arise processes which will try to return the frequency to its former value. Such a system can be called absolutely or ideally elastic. All its parts are mutually dependent. This dependence is realized either directly or through other parts of the system, acting as intermediate links.

As soon as this ideally elastic system (or body) is set in motion with uniform velocity, the equilibrium positions of all its parts will be altered and it will undergo the Lorentz contraction. There occurs also a decrease of the equilibrium frequencies of oscillations or rotations of separate parts of the system. An important role is played by the relativistic growth of mass. To provide the equilibrium of the accelerated system, the parts of the system must acquire not only new frequencies of their oscillations or rotations, but also become late or fast relative to each other in accordance with their location along the motion. Within the system or inside the body there appear time zones. That's why any perfectly elastic system, as soon as its velocity changes, behaves in full accordance with the Lorentz transformations — the dimensions of interacting bodies as well as the distances between their parts reduce in the direction of motion, the rates of all the processes are decreased; the processes going on in the front parts of the system or of the body prove to be late as compared with the processes taking place in the rear parts of the system. If there is an observer inside the system, he or she can feel the change in velocity only during the process of acceleration, or some time after it. This time is required for the parts of the system or the body to find new positions of equilibrium for themselves, to acquire new frequencies of oscillations or rotations, and also for the creation or rearrangement of the time zones. After this transient is over, the observer again sees the system the way it was before the change in the velocity, all the properties of the system being restored. If the observer overslept the process of acceleration as well as the transient, and his instruments memorized nothing, he would then see all processes inside the system proceed further just in the way as if the velocity had not changed. It would be only the foreign instruments that could register all the alterations inside the system, provided the velocity of those instruments had not undergone any change.

Not so for a system whose parts are isolated of and cannot interact with each other. These parts have neither equilibrium positions nor local equilibrium oscillations or rotations. When the uniform velocity of such system changes, the relative positions and velocities

2.7. EINSTEIN'S POSTULATES   183

of its isolated parts may experience any change. In particular, they may not change at all if the isolated parts were accelerated in identical conditions. We have already mentioned two rods, or, even better, two halves of a broken rod, being accelerated identically. Each half contracts its length by a factor of γ, but the distance between the centers of the halves remains unchanged. As for the gap between the halves, it becomes even greater. We have also considered the two spatially separated clocks which were synchronized with light signals and then stopped communicating with each other. If these clocks, that are absolutely independent of each other, undergo identical acceleration, the changes in their readings, if any, will be also identical. After their transfer into the moving world, they will prove non-synchronous. If in the moving world these clocks are slowly brought together, their hands will find themselves in different positions and this difference can be used to estimate the change in the velocity of the system — even if the observer overslept the acceleration with his instruments switched off. A similar estimation can be also made by measuring the distance between the centers of the two halves of a broken rod. For the observer who has undergone acceleration, this distance increases by a factor of γ (due to the contraction of his tape-measure), and he will be able to use that increase in order to determine γ, and thus to figure out the change in the speed v without going beyond the limits of his system.

It should be stressed that with the change in velocity, every separate clock changes its tick, because each separate clock is an equilibrium elastic system. But the clocks, not knowing anything of each other, change their readings identically. They behave like the two halves of a broken rod. To use a broken rod after the acceleration, its two halves must be glued together. For the same reason the independent clocks must be synchronized after their acceleration. Otherwise they will not be able to perform their functions.

For the two clocks to work like a whole rod, they must be continuously and firmly synchronized with each other. Let clock A be an independent generator of time marks, which are received and repeated by clock B at some other point of space. Such clocks constitute a united, absolutely elastic system, and an increase in their speed not only causes a slowdown of their tick, but also makes the front clock (in the direction of motion) slow with respect to the clock located behind. Detailed reasons for that delay were discussed in Section 2.4. They were connected not with the overload, arising in the course of the acceleration of the clocks, but with the electromagnetic wave by which the clocks communicated with each other. When these clocks change the velocity of their motion under the action of a non-gravitational force, it does not tell on the wave. Thus, for a certain period of time, the crests of the wave are arriving at the slave clock either more or less often, in accordance with its position relative to the master clock (behind it or in front of it with respect to the direction of motion)

2.7.5. Crocodile scenario

To have a better feeling of a relativistic elasticity, let us imagine a huge female-crocodile (an enormous monster from a fiction novel, outlined in Fig.21) with a newborn baby crocodile playing gaily near its tail. Both of them are located on a huge platform which initially is ar rest. Assume that the platform with the crocodiles is set in motion, the mother's tail forward, at a uniform speed close to that of light. At one short moment of acceleration, the two creatures will suffer enormous overloads which would be mortal for any living being. But we will disregard the overloads as we did earlier with respect to the pair of clocks. Let our fantastic monsters endure very high, though very short, acceleration with all the particles of their

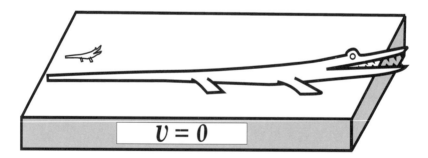

Fig.21. A female crocodile of fabulous length with its newborn baby located at the end of its mother's tail. Sitting on the platform, the family is ready for being set in motion at a uniform speed close to that of light. Both the creatures are painted white to indicate zero increments of their ages with respect to the moment of setting them in motion.

2.7. Einstein's postulates

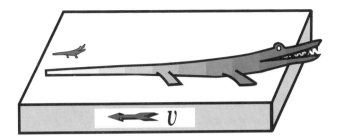

Fig.22. The female crocodile of length l with its newborn baby (shown in Fig.21) have been set in motion at a uniform speed, tails forward, together with their platform. Being very long, the mother has needed as much time as more than l/c to recover from the fast acceleration. This time is necessary to the particles in the mother's body to find their new equilibrium positions – closer to each other than they were before the acceleration. The body of the mother has not only experienced the Lorentz contraction, but also acquired time zones distributed along its length, because the electromagnetic field between the particles of the body did not take part in the common acceleration. The end of the mother's tail is now younger than her head by $(xv)/c^2$. These time zones are indicated by halftones of grey. The older the part of mother's body, the darker its shade. As for the baby, it is an independent being, whose parts of the body are not connected to those of its mother's in any way. Therefore it has been aging at the same rate as its mother's head and, hence, faster than the mother's tail.

bodies left in their former positions. When the platform was at rest, these particles were in equilibrium, which is now violated by the motion. Now, these particles will have to find new positions of equilibrium for themselves. This will take a lot of time for the mother, who is so long that even light seems slow in providing communication between different parts of her body. This time must be at least l/c, where l is the length of the body and c is the speed of light. After the time of the transient has passed, the length of the mother will be contracted γ times and the time zones will be formed along the mother's body just in the same way as described in Section 2.4 for the case of a pair of spatially separated clocks continuously synchronized with each other. (See Fig.18 on page 142 with explanation given there.) The mother's head will then prove to be in one time zone, while her tail and the baby will be in another time zone, which

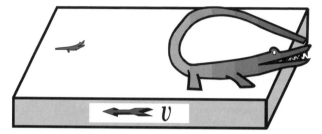

Fig.23. The mother-crocodile is inspecting her tail. Before inspection, the end of the tail was younger than the head. But in the process of bending, the end of the tail has been moving a bit more slowly than the platform and, therefore, was ageing faster than the head, which has been fixed to the platform. The head and the end of the tail, when they meet each other, prove to be of absolutely the same age, so that the crocodile is unable to notice that many parts of her body became younger than her head as a result of the change in velocity. After being returned to its former place, the end of the tail will again become younger than the head, but the crocodile will not be aware of it. As for the baby, it became older, but the difference in age between it and the head of its mother remained the same. We wonder what will happen if the baby takes a ride to the head of its mother on the tip of the mother's tail. During the ride, it perhaps would be aging with the same rate as the end of the tail. But do you remember that before the ride it already was older than the tail?

is reflected in Fig.22 by the halftones of grey. The crocodile's tail, connected tightly to her head through many different signals, becomes younger than her head in accordance with the time zone where it happens to be. (For clarity we assume that the mother's master-clock is located in her brain.)

The halftones in Fig.22 indicate the actual time of existence of the baby and of different parts of the mother's body as counted from the moment of the acceleration. This time is the shortest for the tip of the mother's tail, painted white, and the longest for the mother's head (and for the baby as well) painted dark-grey. The baby is regarded short enough to neglect the difference in these times between the parts of its body. The baby is darker than the mother's tail because, unlike the tail, it is an independent entity which lives and ages in the same rate as the mother's head.

2.7. EINSTEIN'S POSTULATES

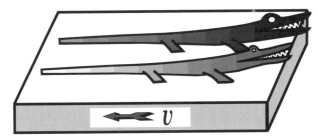

Fig.24. This situation has arisen when the baby-crocodile made a slow travel along his mother's body (or was transported by the mother on the end of her tail) until their heads were alongside each other. At the start of this travel the baby was of the same age as his mother's head (and older than the mother's tail). But during the travel, the baby was moving a bit more slowly than the platform and, therefore, was aging faster than its mother. The travel proved long enough for the baby to grow up and even to become older than its mother.

The mother-crocodile is unaware of the difference in age between the parts of its body. If, for example, she bent her tail and brought the end of it to her head (as shown in Fig.23), then after smelling and examining it carefully, she would be able to make sure that her tail is of the same age as her head. This is because in the course of the bending, the end of the tail is aging faster than the head, so that the moving observer (in our case the mother-crocodile) is not able to detect her motion by inertia.

Now, let us see what will happen if the baby takes a ride to the mother's head on the tip of the mother's tail. During that ride it will be aging (due to its slow transportation) at the same rate as the tip of the tail, so that by the end of the ride, it will become older than not only the mother's tail, but even her head as shown in Fig.24. That is how amusingly space and time are entangled when spatially separated objects, independent of each other, change the velocity of their motion. We must remember, however, that traveling through time may be realized in only one direction − in the direction of our future. As for our past it is impossible for us to get there. We may get, though, into the past of another person, but only on one special condition: that man must have always been so far from us, that we could not have ever heard anything from him, even by radio.

2.7.6. Clock paradox

Another example. Two twin brothers Mike and Nick were sent to the depths of cosmos in order to see how motion with a uniform velocity affects the relative age of human beings. There are two huge platforms at their disposal – specially equipped for their measurements. At first they are stationed on one of them. They call this platform "home platform". It is regarded as being at rest. The brothers know that, according to their measurements, the length l of the home platform is equal to 800 light-days. The other platform moves parallel to home platform with a uniform speed $v = 0.99875c$ as shown in Fig.25. This speed corresponds to the value $\gamma = 20$. The experiment begins when the front edge of the speeding platform comes alongside the brothers and Nick jumps onto the speeding platform.[1] Mike remains on the home platform. The start disposition of the brothers is reflected in Fig.25. Mike and Nick are continually tracking the ages of each other. When doing so, each of them trusts only his own clocks. To perform this tracking, Mike has placed his stationary automatic observation posts along the whole path of Nick's travel. Each post is supplied with a clock-calendar and a high-speed camera triggered automatically. When Nick passes by each post, the camera shoots and makes a snapshot that simultaneously shows both Nick's personal clock in motion and a stationary observation post with Mike's clock attached to it. The reading of Nick's clock in the picture shows the time that Nick has actually lived since his start. The reading of Mike's clock in the same picture shows the time that Mike has lived since the moment of start according to his own opinion. Mike's clocks have all been, in advance, synchronized with light signals according to Einstein, or by means of slow transportation.

Nick mirrors Mike's measurements. His observation posts have been placed on the speeding platform before the start. They are of the same design and arranged along the platform in the same way as Mike's stationary posts. When such post speeds past Mike, the moving camera shoots and creates a snap in which we can see Mike's personal clock and the relevant clock attached to the moving post. The

---

[1]. As usually, we do not care about the enormous overloads experienced by the traveller while jumping onto the platform. To stay within the scope of special relativity, we just assume that Nick somehow remains safe and sound and in a short period of time undergoes the Lorentz contraction and other changes inside his body predicted by special relativity.

2.7. Einstein's postulates

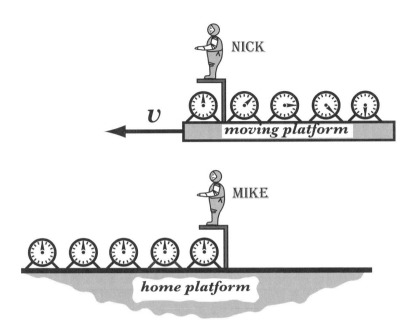

Fig.25. Two twin brothers Nick and Mike at the start of their space-time experiment. Nick starts on 800-days space travel on a long platform with a uniform velocity v, while Mike stays on the home platform and observes his brother through a lot of observation posts distributed along the path of his brother. Each post is supplied with a camera which shoots automatically as soon as Nick with his personal clock comes alongside the post. Nick has a similar set of observation posts with cameras, distributed behind him along the platform. Each camera will shoot as soon as Mike comes alongside the post.

interpretation of the clocks' readings is similar. Mike's clock shows the time that Mike has actually lived since the start. Nick's distant clock shows the time that Nick has lived since the departure according to his own opinion. All his spatially separated clocks have been synchronized in advance. At the moment when Nick jumps onto the platform by his first observation post, the clock at that post shows 12 a.m. o'clock, January 1, 2010. Nick's other observation posts are behind Nick at that moment, as shown in Fig.25. All Nick's clocks have been synchronized according to Einstein and are thus in different time zones. The further from Nick, the faster the clock. The very last clock is fast as compared with Nick's personal clock by a value of $\beta\gamma l/c = 16,000$ days, or by approximately 44 years, so that at the moment of the start it shows October 2, 2053. That number of posts would have been sufficient if Nick's travel lasted even as long as almost 44 years by his clocks. In reality, as we will see, Nick's travel will prove $\gamma^2 \cong 400$ times shorter, so most of his observation posts will be spared. Not only the number of the moving posts, but also the length of the platform could be considerably smaller, but Mike and Nick are unaware of it and have built such a long construction just in case.

The travel starts at the very beginning of year 2010 and lasts 800 days by Mike's clocks. By the end of the way Mike's clocks show 800 days (March 10, 2012). For Nick and his clocks time passes 20 times more slowly, so by the end of the travel Nick's personal clock reads $800/\gamma = 40$ days (February 9, 2010). From the beginning of the travel to its end, Mike lives more than 2 years, and Nick – only 40 days. Nick proved to be younger than Mike by 760 days, which is over 2 years. This is what Mike learned from his snapshots, taken at his observation posts. It's very interesting what Nick thinks about it.

When passing by Mike's distant observation posts, Nick sees the readings of Mike's distant clocks fixed there. But he does not make use of them for two reasons. First, the brothers agreed before the travel that each of them would use only his own clocks. Sharing Mike's clocks would mean a violation of their convention. Secondly, Nick has a right to regard his own clock as being more reliable than Mike's, for Nick assumes himself to be at rest, and he assumes Mike to be in motion. He heard that motion affects the clocks and hence regards Mike's clocks as unreliable. His idea of Mike's age is formed on account of the snapshots made from the observation posts speeding past Mike. Suppose Nick wants to see how long Mike has actually lived since the start. Remember that Mike himself regards this period as equal to 800 days. Nick does not care what Mike and Mike's instruments think about it because he does not trust them. He wants to see what has been registered on his own, moving observation posts. By the end of the travel Nick's personal clock shows 40 days (February 9, 2010). Thus Nick must look through all the snapshots of Mike made from Nick's moving observation posts and find among them the one in which the moving (i.e. one of Nick's) clock shows February 9, 2010. Mike's personal clock will also be seen in that picture. We wonder what will it show?

The previous section tells us that it will show the time which is $\gamma = 20$ times shorter than that shown by Nick's personal clock. It is as short as 2 days only. Nick's clock in the picture will show February

9, 2010, while Mike's clock will read 12:00 a.m., January 3, 2010. This snapshot, in Mike's opinion, was made as early as two days after the start, when Nick had made only 1/400th part of his travel. But according to Nick, this snapshot was made 40 days after the start, i.e. after he has covered the entire distance.

Thus, Mike and Nick, on account of their own clocks, have different opinions of each other's age at the end of the travel. Mike is sure that he himself has lived 800 days and that Nick has lived only 40 days. Nick is sure that he himself has lived 40 days and that Mike has lived only 2 days. Each of them regards himself as being older than his brother at the end of the travel. Mike thinks he is older than Nick by 760 days, and Nick thinks he became 38 days older than Mike. This is in full agreement with special relativity. Though the brothers' statements are opposite to each other, both of them are right. Neither of them can be disproved as long as Nick continues to move with a uniform velocity. If he keeps his motion forever, no one will ever be able to determine whose opinion is right – Mike's or Nick's, just as it is impossible to say which of the two brothers is in motion and which of them is at rest. This will be true indeed because in this situation the brothers will never meet and will have no chance to see with their own eyes which of them has been right or wrong.

But our experiment will have another development. At the end of the distance, Nick quickly jumps down from the speeding platform onto the home platform together with his personal clock and brings himself to rest. We assume that while he is jumping down nothing substantially affects either him or his clock. His personal clock confirms this assumption – it shows the same time as it did on the platform – February 9, 2010. All the observation posts of Nick's might have jumped down from the platform together with Nick and might have also acquired a state of rest. Even if they all remained safe and sound, they would become non-synchronous and all their measurements would be wrong. Nick cannot use them any more. His voyage is over. He is now on the home platform – free from his obligation to use only his own spatially separated clocks. Now, he can use Mike's observation posts, including the one of them located nearby. All of them, including Mike's personal clock show the same time – March 10, 2012.

We see that Nick's opinion has suffered a great change. When on the platform, he thought that Mike's age had increased by two days since the start. After jumping, he thinks that Mike became as much as 800 days older during the time of the travel. As a result of Nick's jumping down from the platform, Mike, in Nick's opinion, became older by as much as 798 days.

Before Nick jumped down from the platform, he thought that Mike was 38 days younger than Nick. Now he sees Mike become 760 days older than Nick. What is it that has really changed — Mike's age or Nick's idea of Mike's age? Nick's idea, of course! Because nothing could have happened to Mike due to the fact that Nick had jumped down from the platform. And nothing has happened to Nick either. The only thing that has changed is the choice of spatially separated clocks used by Nick in his estimation of Mike's age.

Now we have to bring the experiment to an end. Nick comes back to Mike with a low speed (not higher than, say, $c/10$). The speed being low enough, the relativistic effects may be neglected. Thus, when Mike and Nick meet in 20 years or so, both of them see with their own eyes that Mike is older than Nick by 760 days. This is also proved by the readings of their personal clocks, that either of them has always kept by him. The same refers to the nuclei of iron in the experiment, described in Section 2.3. After all it was Mike's instruments that proved to be right, because they had never changed their velocity during the experiment. Mike's frame of reference was remaining inertial in the course of the experiment, while that of Nick's was not.

2.7.7. When non-postulated relativity brings about the result much sooner than the traditional theory

We can also consider one more example, taken from radio engineering. Exactly above the radar there is a fast flying rocket. The radio signal of the radar is reflected from the rocket and received on the surface of the Earth. Will the motion of the rocket affect the frequency of the reflected signal? Try to answer this question by yourself. The answer is no, it will not. The answer would be valid even if the speed of the rocket were close to the speed of light. When the radio signal is being reflected from the rocket's airframe, there appear currents in the airframe, whose frequency is equal to the frequency of the incident signal. It is these currents that excite the reflected radio signal. If these currents were produced by the rocket oscillator, then, due to the motion of the rocket, their frequency would be lower by a factor of γ. But in the case involved the currents

2.7. EINSTEIN'S POSTULATES

are driven not by a local master generator whose frequency depends on the speed of the rocket, but by an external radio signal, that arrives perpendicularly to the motion. The currents are not free. They must oscillate with the frequency of the incident signal that excites them. Even relativity cannot disprove this fact. Thus, in full accordance with relativity, the frequency of the reflected signal is equal to the frequency of the incident wave. The alternating currents in the rocket airframe are a kind of a clock. But this clock is not independent. It is strongly synchronized with the radar generator. If there was a retransmitter of the coming signal on the rocket, everything would be the same. The confusing circumstance is that one of the two synchronized clocks is at rest and the other one – in motion. Einstein did not consider such situation.

It's very amusing that there was a hot discussion on this problem in science literature of last decades. About half of the authors were of one opinion, the other half – of the other, both parties referring to special relativity. Eventually an experiment was made with a great precision, worthy of the second half of the 20th century, which removed all doubts. It is strange that none of the specialists had even thought of considering or checking out this problem without referring to relativity. Then, most probably, this experiment would have not been needed.

Summing up what has been said in this section, we can state that the most difficult things are already behind. We were able to behold time and space as Einstein saw them and to discover a lot of amazing and interesting things in these seemingly "simple" notions. Actually, we have "revived" four Lorentz transformations: (1.11), (1.12), (1.13) and (1.20) (see page 48), pertaining to space and time. There remains only one step to make the Lorentz transformations completely revived. We have to instill a physical sense into the transformation (1.19) that governs a charge density. This time, we know even in advance that ρ' should be the density of the charge, registered by moving instruments. But why does the expression for it look so odd? We will consider this question in the next section.

2.8. Electrification of currents

where we will see that a neutral current-carrying loop has a good reason to acquire electric polarization after being set in motion with a constant velocity

Let us imagine a charged plane capacitor, consisting of two horizontal plates A and G (see Fig.26). Let this capacitor move with a speed v in the direction $O'O$, parallel to the plates. Inside the capacitor there is an electric field $\mathbf{E}_0$ perpendicular to the plates and pointing upward. The electric charges, stored on the plates of the

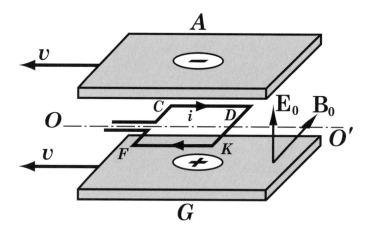

Fig.26. The plates A and G comprise a plane capacitor, which is moving from the right to the left with a uniform velocity v. Since the charged capacitor is in motion, the space between the plates is filled with not only the electric field $\mathbf{E}_0$, but also the magnetic field $\mathbf{B}_0$. The current-currying loop $CDKF$ (being either in motion or at rest) is placed between the plates to detect this magnetic field by its rotation round its axis OO'.

2.8. ELECTRIFICATION OF CURRENTS 195

capacitor, move together with it. But the charges, when moving, make currents, which, in their turn, always produce a magnetic field of a certain induction $\mathbf{B}_0$. This field is effective in the space between the plates. It points in horizontal direction and is perpendicular to the line of motion $O'O$, as shown in Fig.26. The magnetic induction B_0 is proportional to the speed of motion β:

$$B_0 = \beta E_0.$$

If the capacitor were at rest, its velocity would be equal to $\beta = 0$, and there would be no magnetic field anywhere. With the instruments fixed to the capacitor, is it possible to establish the presence or absence of the magnetic field, and thus distinguish a moving capacitor from the stationary one? In other words, can the observer fixed to the capacitor hunt down the ether drift? If a device could be found that would respond only to a magnetic field and to nothing else, then such device could be placed between the plates of the capacitor to give an output proportional to the speed of the ether drift.

At the first sight, it seems that such device could be designed without any problems. It seems enough to take an ordinary compass, place it between the plates, and watch the deviation of its pointer. If the pointer turns, then there is a magnetic field there, which will mean that the capacitor is in motion. In the opposite case there will be no magnetic field, which will mean that the capacitor is at rest. No reasons are seen on the face of it to prevent the pointer from turning in the magnetic field. This argumentation seems irrefutable: the electric field, whatever it may be, seems to be unable to turn the pointer, and there is nothing else between the two plates but the electric and the magnetic fields.

To clarify the situation, let us replace the magnetized pointer with a current-carrying rectangular frame $CDKF$ shown in Fig.26. There is no principal difference between such frame and the pointer. Both of them must respond to the magnetic field in the same way. The electrons moving along the conductors CD and KF are acted upon by a Lorentz force, perpendicular both to the currents in the conductors and to the magnetic field. As a result, the conductors CD and KF must be pushed upward and downward, respectively. Under the action of these two forces the frame is compelled to turn round its axis $O'O$ and take a vertical position, so that its plane will be perpendicular to the induction $\mathbf{B}_0$. There are a lot of such microscopic frames inside a magnetic pointer. It is they that make the pointer turn toward the direction of the magnetic field. It will be sufficient for us to consider just one of such frames.

To aggravate the situation let us modify the problem. Let our capacitor move to the left at a speed v, while the frame remains at rest. Then the frame is sure to turn. If the frame had not turned, no electric motor would be ever able to work, because the only difference between the frame and the rotor of a working electric motor is that the frame has no collector. Had we attached a collector to it, it would rotate all the time. It would even perform work at the expense of the energy supplied by the source of the current driven through the frame. This is a normal situation with no traps hidden or questions arisen.

Now, let us reverse our scenario. Let the capacitor be at rest with a current-carrying frame moving through it from the left to the right at a speed v. The behavior of the frame must be obviously the same as in the previous case. The frame must turn if it hasn't got a collector, and if it had a collector, it would rotate continuously. It does not matter which of them is in motion and which is at rest – the frame or the capacitor! It is just the way how events occur according to Einstein's first postulate. But what turns the frame now, when the capacitor is at rest and there is no magnetic field in it except the field produced by the frame itself? But the frame cannot turn itself just by means of its own field. Like a man cannot lift himself by his bootstraps. Or maybe it is Einstein's first postulate that makes the frame turn? But the postulate never does such things by itself. It does it only through relevant laws of nature, that serve it loyally and follow it obediently. We have had a lot of chances to make sure of it in the previous sections. When a miracle was needed to justify the first postulate, the miracle would take place, and it would prove eventually that all miraculous things have quite a natural explanation following from well-known laws of mechanics or electrodynamics. Something like this must happen now.

The clue is hidden in the frame $CDKF$. Let us make a separate picture of it (see Fig.27). Let us focus our attention on the conductor DK. There is a current in this conductor, which creates a magnetic field **B** around it. This field propagates together with the moving frame. The magnetic lines of this field have the shape of rings strung

2.8. ELECTRIFICATION OF CURRENTS

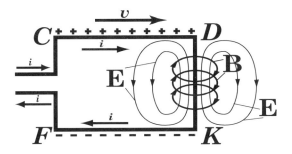

Fig.27. This outline shows how a current-carrying loop $CDKF$, moving from the left to the right at a uniform velocity v, acquires an electric polarization. The magnetic field **B**, produced by the current i and transported together with the loop, generates the vortical electric field **E**, which follows the loop and drives the electrons along the conductor DK, so that the conductors CD and FK become and are kept oppositely charged.

onto the conductor DK. Similar rings arise around the other conductors as well, but for clarity they are not shown in the figure.

The frame moves with a uniform speed v from the left to the right. Imagine we are at a certain stationary point Q on the path of the frame. The frame speeds past us and we see that the magnetic induction **B** at the point Q at first is growing (as the conductor DK approaches us), and then abruptly changes its polarity (when DK passes through the point Q), after which **B** decreases and vanishes. In short, the induction **B** at the point Q changes with time. And we know what must happen when a magnetic field changes with time. The law of electromagnetic induction begins to work and gives rise to a vortical electric field **E**. The lines of force of this field have also the shape of closed loops, that are strung onto the magnetic lines as shown in Fig.27. In the conductor DK this field is directed from the point K to the point D. Under the action of this field the electrons inside DK rush to K (electrons being charged negatively) and spread along the conductor KF. Their accumulation there produces an electric counter-field (not shown in Fig.27 for the sake of clarity) that acts from point D to point K inside the conductor DK. Ultimately the net electric field within all the conductors of the frame will become zero and the whole process will come to an end with a steady excess of electrons in the conductor KF, and their according deficiency in CD. Similar processes will take place in the conductors CF and close to them, which will be in accord with the processes which take place in the conductor KD.

It is clear now what it is that makes the moving frame turn inside the stationary capacitor without any external magnetic field. The frame turns under the action of the electric field $\mathbf{E}_0$ of the capacitor. The conductor KF is charged negatively and is attracted to the plate G, while the conductor CD, charged positively, is attracted to the plate A. The torque, when calculated, proves to be the same as in the case of a moving capacitor, whose magnetic field $\mathbf{B}_0$ acts on the stationary and thus non-electrified conductors of the current-carrying frame.

The electrification of a current-carrying frame takes place even when the frame and the capacitor move together. The frame is then acted upon by two torques, equal in magnitude and opposite in direction. The first torque is caused by the magnetic field inside the capacitor. Due to the current of electrons, taking place within the conductors CD and KF, there arise two opposite Lorentz forces that are exerted on these conductors and make a torque exerted on the frame. The second torque is caused by the electrification of CD and KF and their electric attraction to the charged plates of the capacitor. These two torques cancel each other and the frame does not turn. The ether drift again proves imperceptible. An observer, moving together with the frame, will give, as usual, his own explanation of the events. According to him, there is no magnetic field in the capacitor, and so, there are no Lorentz forces there. Neither there is any electrification of the frame. Thus there are no reasons for the frame to turn. And the frame does not turn indeed, which confirms the reasoning of the moving observer. But our reasoning, based on the magnetic field and the electrification, brings us to the same conclusion, and therefore, is also correct. It is not for the first time that we face such a situation.

You might be puzzled. How can it happen that the observer who is moving together with the frame has no way to see that the conductor KF contains more electrons than the conductor CD? If going to all lengths, what prevents him from counting those electrons just directly? And if he succeeded in doing it and discovered the electrification, he would immediately become aware of his motion with uniform velocity. By changing that velocity, he would eventually come

2.8. ELECTRIFICATION OF CURRENTS

across such a speed that would not result in any electrification, and would correspond to his being fixed to the ether. Well, let us see how he or she will cope with counting these electrons. It is not so simple as it seems at first sight. Not because they are too numerous, but because they are moving along the conductors of the frame. One has to count the geese in a running flock.

To simplify the task, we will assume that the electrons are running in single file. The observer is to find out how many electrons are in the conductor *KF simultaneously*. To do it, it is necessary to mark the first and the last electrons of those to be counted simultaneously. But that cannot be done without the clocks, synchronized to each other and installed at points K and F. Because the frame is moving, the clock K will be late with respect to the clock F. Suppose the automatic devices, marking the electrons at points K and F, are adjusted in a way that the mark must be made exactly at 12 o'clock at both points. The electrons run inside KF from the left to the right, i.e. in the direction, opposite to the electric current i (see Fig.27). When the device at point F marks the electron passing by it, the device at point K is still waiting for its time to come. The electrons leave the conductor KF for the conductor KD, turning round the corner K, while the device at the corner K is still waiting for its time to act. It goes without saying that with such a "non-synchronous" work of the two devices, the number of electrons in the conductor KF, counted by the moving observer, will prove to be less than the number of the electrons that simultaneously occupy KF according to the observer who is at rest. Eventually, it will turn out that according to the observer who is in motion, there is no excess of electrons in the conductor KF or, in other words, there is no electrification there. The moving observer will not notice any excess of electrons, should he even count them by finger, provided this method "by finger" is synchronized according to Einstein.

The electrification of a moving current-carrying frame can be also explained in quite a different way based on mechanics rather than electrodynamics.

The electric current in a moving frame is caused by the motion of the electrons with respect to the frame. Therefore the electrons in the conductor KF are moving a bit faster than the frame, while in the conductor CD their net speed of motion is a bit lower than that of the frame. Because the mass of the electrons depends on the net speed of their motion, the mass of the electrons in the conductor KF is a bit greater than in CD. When any of these electrons, participating in the current, leaves the conductor KF for the conductor CD, its mass becomes a bit smaller. According to the law of conservation of momentum, this tiny decrease in the mass causes a corresponding small increase in the electron's velocity with respect to the frame. This means that the stream of the electrons in CD is flowing a bit faster than in KF. But the stream is continuous, it never ceases. So, if a certain number of electrons leaves some place for the downstream regions, the same number of electrons must arrive there from the upstream regions. The electrons drift just like a river, at some places faster (the conductor CD), in other places – more slowly (the conductor KF). Suppose we divide the river into equal lengths. Which of them will contain more water – the one with a slower or the one with a faster stream? It is common knowledge that the slower and calmer the river flows, the wider it becomes. That's why the number of electrons in the conductor KF will be greater than in CD. This reasoning can be supported by a relevant calculation. The result is the same as in the case of the explanation given in terms of electromagnetic induction. It is remarkable that this phenomenon can be explained equally well by means of such different parts of physics as mechanics and electrodynamics. This suggests a deep internal connection between them, while relativity acts as a connecting link.

The electrification of a moving current-carrying frame is a particular case of the following general rule that follows from relativity:

When a current-carrying loop moves with a uniform velocity in a certain direction, then the parts of the loop are charged either positively, when the current is directed along the translational motion of the loop, or negatively, when the current is directed against the motion. As for the net charge within the loop, it always remains the same.

There are two velocities which determine the amount of the additional charge, piled up in the upper and lower conductors of the rectangular frame in the course of its electrification. The first velocity v refers to the frame as a whole, while the second velocity u specifies the motion of the electrons within the conductors of the frame.

2.8. ELECTRIFICATION OF CURRENTS

Suppose the velocities v and u are given, and so is the density ρ' of the charge within a stationary conductor. As soon as the current-carrying conductor starts its longitudinal motion with a constant speed v, the density of the charges drifting inside it either increases or decreases depending on whether the directions of the current and v are the same or opposite. Therefore the charge density ρ in the moving conductor differs from ρ' in accordance with the following relation:

$$\rho = \frac{\rho'}{\gamma \left[1 - \frac{uv}{c^2} \right]} . \qquad (2.53)$$

If the current and v have the same direction, then the charge density ρ is greater than ρ'. In the opposite case motion results in the decrease of the amount of charge in the conductor.

Comparing the relation (2.53) with the Lorentz transformation (1.19) on page 48, we can see that these two relations just coincide (with taking into account that, in our explanations, the motion is always directed along the x-axis, so that $u = u_x$). But this time the charge density ρ' is not a fictitious variable in a fairy world, but the density of the charge in the stationary current-carrying conductor. We have the right also to say that ρ' is the charge density in the moving current-carrying conductor as measured by the instruments moving together with the conductor. This will be true because these instruments are "unaware" of their motion.

The relation (2.53) can be derived in three different ways, which were used above. The first of them is based on the law of electromagnetic induction, the second – on the law of conservation of momentum, and the third – on special relativity. When using the third method, it is sufficient to require that the frame with a current, moving together with the capacitor, should be indifferent to the presence of the magnetic field (otherwise the principle of relativity would be violated). Technically this method is the simplest, though it does not expose the actual cause of the electrification of the frame.

When we use the rule of the electrification of a moving current, some precautions have to be taken. The following conditions should be satisfied: the motion of the current-carrying conductor must take place at a uniform velocity, and not only the separate conductors, but the whole closed electric circuit comprising these conductors must take part in the motion. If these two conditions hold, the rule of the electrification of currents is always observed. Otherwise this rule may be violated.

Let us consider for example the electric circuit $ABCD$, shown in Fig.28 (a), with a current i, driven through it. The conductors AB, AD and DC are at rest, while the conductor BC is part of a long wire, connected with our circuit by means of the sliding contacts B and C. The wire BC is moving at a uniform velocity $\mathbf{v}$. This wire is not electrified despite the electrons of the current i are dragged by the moving wire. The transverse conductors AB and DC are at rest, so that the law of electromagnetic induction does not work. Electrification of the wire will not take

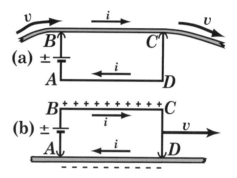

Fig.28. (a) An example of a moving current which is **not** electrified in spite of its motion. The moving current-carrying wire is sliding over the brushes B and C, while the other parts of the circuit are at rest. It is in contrast to the case (b), where the moving part of the circuit does acquire the electrification. Even Einstein's postulates, if taken alone, prove helpless in solving this puzzle.

place just because there isn't any ground for it. This scenario might be compared with another case, shown in Fig.28 (b). The conductor BC is moving there together with the transverse conductors AB and CD, sliding along the stationary wire AD. It's here that the formula (2.53) works with all its might. The law of electromagnetic

2.8. ELECTRIFICATION OF CURRENTS

induction would work there as though the wire AD were also in the motion. So not only the moving wire BC but even the stationary conductor AD would be electrified. It cannot be otherwise. If AD were not electrified, where would the excessive charge in the wire BC have been brought from? The law of conservation of charge is not ever violated, is it? It should be noted that the system shown in Fig.28 (b) is nothing else but the system shown in Fig.28 (a) as seen by an observer who moves together with the long wire BC. Do you see how different one and the same system looks when it is examined by two different observers, one of them being at rest and the other – in motion. Perhaps it doesn't surprise you any longer.

A lot of misunderstanding and bewilderment is sometimes caused by the following "very simple" problem. Assume that there is a metallic rim with a circular current in it. We put the rim on a wheel and set it rotating. There is not only a current but also the motion of the conductor along the current. It seems that the rim must get charged immediately. But where from can it get the charge? In the case of a moving circuit or frame there is no need in any additional charge. The charge is only redistributed between different parts of the system. But in the case of the rim, the charge is apparently only growing without any compensation. Because the law of conservation of charge cannot be disobeyed, the rim will not get charged at all. And what about the relativity postulate? Isn't it violated here? No, it isn't. Special relativity can be applied unreservedly only to a motion with a uniform velocity. And we have a rotational motion there, which is a special kind of accelerated motion. Very often, though, special relativity proves applicable (even with good accuracy) to rotating electrical devices such as for instance electric motors. Some examples of this will be given below.

If a stationary charge q is placed between the poles of a stationary magnet, the charge will not of course be acted upon on the part of the magnetic field. But if the magnet and the charge are set in a common motion with a uniform velocity **v** in the direction perpendicular to the vector of the induction **B**, then the charge will be acted upon by the Lorentz force $q\beta B$. If under the action of this force the charge q was displaced relative to the magnet, then the observer who moved together with the magnet could immediately detect his motion and register the ether drift. But nothing of the kind takes place there, because the winding of the moving magnet is electrified, which gives rise to the electric field $E = -\beta B$ that completely counterbalances the action of the magnetic field upon the moving charge. The observer in motion does not notice the electrification, but according to him, the charge q is at rest ($\beta = 0$) so that the Lorentz force $q\beta B$ does not arise. If the magnet in motion is a permanent magnet and has no winding, it will be electrified all the same. The magnetic properties of such magnet are caused by microscopic currents in its atoms and even deeper. Those currents also obey special relativity and, when set in motion, are sure to be electrified. Formula (2.53) and the rule $E = -\beta B$ always hold when the velocity of motion is uniform.

T he relativistic effects, discussed in this section, are remarkable for being apparent even with a speed of motion far from the speed of light, for instance with the speed of rotation of an ordinary electric motor. Let us consider for example the work of a direct current generator, whose operation is illustrated by Fig.29. The permanent magnet has a form of a solid conducting cylinder magnetized along its axis. The magnet is set in rotation around its axis OO' with a usual for the electric generators rate (say, 1500 revolutions per minute). There appears a voltage between the stationary

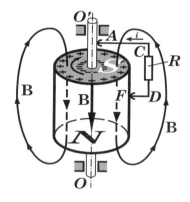

Fig.29. Producing of electric current i in a stationary external circuit $ACDF$ by means of a rotating conducting cylinder magnetized along its axis OO' and rotating around that axis. When the brushes A and F are disconnected from the cylinder, where, then, will the electric charges pile up – on the brushes or on the cylinder? If you can answer this question, you understand how a unipolar generator works.

brushes A and F, and there arises an electric current i, sent through a stationary resistor R. How and where does the electromotive force, driving the current, arise?

2.8. ELECTRIFICATION OF CURRENTS

The electric circuit consists of two parts: a moving part, comprising a rotating magnet, and a stationary part, including brushes A and F, connecting wires $ACDF$, and a resistor R imitating the consumers of the generated power. Let us mentally remove all the stationary parts, leaving only the rotating magnet. There is a Lorentz force βB inside the magnet, that tries to shift the free electrons in the radial direction. The magnet is metallic and there is a lot of free electrons inside it. But none of them makes any attempt to change its place in spite of the action of the Lorentz force. Their indifference is due to the electrification of the micro-currents inside the magnet, that produce a magnetic field. These microscopic currents, as well as their electrification caused by the motion, are of a rather complex nature not to be considered here. It will be enough for us to mention that, like any other real objects in nature, they obey special relativity. Each micro-current may have its own good reason for behaving the way it does, but the result is always the same: there arises an electrification in accordance with the equation (2.53). This gives rise to a radial electric field $E = -\beta B$, that completely balances the Lorentz force. The leaves of an electroscope, attached to any place on the surface of the magnet, will not diverge. That means that unlike the rotor of an ordinary electric generator, the rotating magnet does ***not*** bring about any electromotive force.

An observer, rotating together with the magnet, explains this in his own terms. He will say that there is no electrification there, that the electric field does not arise there either and that a Lorentz force cannot arise there by any means, because, according to him, the electrons within the magnet are at rest. An electromotive force is not generated there just because there is nothing to give rise to it.

Now let the stationary parts of the circuit return to their places, but for a while we will leave a little gap between the brushes A, F and the magnet to prevent the electrons from forming a current. The magnetic field of the induction **B**, as well as the electric field of strength $-\beta B$, exist not only within the magnet, but also outside it, including the stationary connecting wires. The electrons inside the wires being at rest, they are not acted upon by the magnetic field, so the electric field E remains uncompensated. That will make the stationary electrons inside the wires rush to the disconnected brush F and pile there while the excessive positive charge will accumulate in the brush A. If the leaves of the electroscope were attached to any of the brushes, they would diverge, indicating the presence of an excessive charge. If now the brushes are pressed to the magnet, there will appear an electric current in the circuit, caused by the EMF, arising in those parts of the stationary wires that are acted upon by the electric field. The observer, rotating together with the magnet, will again interpret this result in his own way. He will say that the wires $ACDF$ together with the resistor R rotate round the stationary magnet and that this rotation gives rise to an electromotive force, caused by the Lorentz force βB. The opinions of the two observers are coincident in finding the place of the generation of the electromotive force (the external wires), but they differ in interpreting the reasons, causing this force.

If someone ever tells you that relativistic effects are just seeming, for, first, they cannot be noticed by the moving observer and, secondly, they take place only at the velocities close to the speed of light, and are, therefore, of no practical interest, you can refer to this generator (engineers call it a unipolar generator) and ask in your turn if it is sound to regard the phenomenon as "seeming" when it is not only registered by instruments, but also generates an electric power. And yet, if it is not quite clear to you why a relativistic effect goes on so violently at the speed of motion which is million times less than the speed of light, refer then for example to Ampere's law (1.3). The constant of proportionality in the CGS system is equal to $1/c^2$ there. Though it seems "very small", the forces of interaction between the currents prove quite sufficient to set powerful electric motors in motion. If even a weak force acts upon a great number of electrons, it becomes quite sensible. By the way, it is time already to direct our attention to the fact that, strictly speaking, there is no formal border between relativistic and non-relativistic phenomena. The term "relativistic" is usually given just to those phenomena that had not been fully comprehended by physicists before Einstein developed special relativity, though they could discover these phenomena even without relativity, had only Einstein waited a little before creating his theory. On the other hand, such a phenomenon as electromagnetic induction, closely connected with relativity, is never called "relativistic", because it had been discovered well before Einstein. Strictly speaking, all the phenomena, known to us today, must be called "relativistic", because all of them satisfy Einstein's principle of relativity. To make sure of it, it is enough to glance at them from another inertial frame of reference.

2.8. ELECTRIFICATION OF CURRENTS

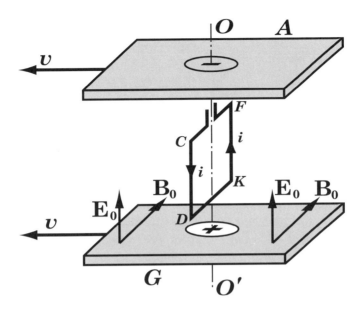

Fig.30. Two plates of a moving capacitor with a current-carrying frame between them. Unlike Fig.26, the axis of the frame is oriented perpendicular to the plates.

Before finishing this section it might be of interest to consider briefly one more group of phenomena connected with the electrification of moving currents. Though these phenomena are unable to cause electrification, they sometimes replace it, when the electrification alone fails to satisfy the principle of relativity. To make it clear, let us revisit the problem of a stationary capacitor with a moving current-carrying frame inside it, that must turn without any magnetic field. In the case shown in Fig.26 (see page 194) the frame turns round the axis OO' due to the electrification of the conductors CD and KF. Let us aggravate the problem by directing the rotational axis OO' of the frame perpendicularly to the plates of the capacitor as shown in Fig.30. If the capacitor is moving and the frame is at rest, the frame will turn round the axis OO' to show the direction of the magnetic field B_0. But what will turn the moving frame when the capacitor is at rest? Whatever the electrification of the conductors of the frame might be, it is unable to make the frame turn round the axis OO', the direction of which coincides with the direction of the electric field $\mathbf{E}_0$ in the capacitor. Because all the forces of the electric interaction between the capacitor and the frame point parallel to OO', they are unable to make a torque turning the frame round that axis. Nevertheless the frame will turn. It will turn according to the third rule of relativistic dynamics (see Section 2.2). Indeed, let us look at Fig.30 and consider any electron inside the conductor CD. This electron participates in two kinds of motion simultaneously. First, it moves in the vertical direction as a participant in the current. Secondly, it moves horizontally together with the frame. The two velocity vectors being imposed, their sum is inclined relative to the direction of the electric field. In this case, according to the third rule of relativistic dynamics, the acceleration gets two components. One of them is vertical and drives the electrons along the conductor. The other is relativistic. It is perpendicular to the conductor and makes the frame turn without any turning force exerted on it. The magnitude of the angular acceleration of the frame is the same as in the case in which the frame is at rest and is turning under the action of the magnetic field of the moving capacitor.

Now that we know the relativistic effects to be significant even at the velocities small as compared with the speed of light, we have realized that, strictly speaking, all the natural phenomena are relativistic. On the one hand, this shows how mature we have become. On the other hand, this is a sign that the book is coming to an end. For it cannot embrace all the phenomena of nature, can it? Here we finish our story about special relativity and switch over to the last section of the book, devoted to the accelerated motion and gravitation.

2.9. The curved emptiness

where, being in search for an inertial frame of reference, we will visit the space and meet free-fallers – the residents of the space – who will explain to us a lot of interesting things about gravitation, which has been thoroughly avoided throughout the previous part of the book

2.9.1. In search for an inertial frame of reference

Einstein's first postulate reads that natural phenomena proceed identically in all inertial frames of reference. But what is meant by those inertial frames and how do they differ from non-inertial ones? How are we to see whether the frame chosen by us is inertial or non-inertial?

The most specific property of inertial frames is based on the fact that all the laws of nature must hold in those frames. Let us take for example Newton's first law, according to which a body, which is at rest or is moving with constant velocities goes on to do so unless some external force is applied to it. Let us see in what frames of reference this law is valid and in what frames it is not. In physics any frame of reference must be associated with a real physical object or a system of the physical objects, linked with each other. The room we are in can serve as an example of such an object. The walls, the floor and the ceiling of the room, that can be extended mentally as far as one likes, form the material frame by means of which we can measure more or less precisely the three coordinates of any event. The clocks, placed everywhere over the volume of the room and synchronized according to Einstein, allow us to determine the time of any event, whatever place of the room it might occur.

Let us place our room somewhere in the depth of the space, far away from significant gravitating masses. Everything in the room will acquire the state of weightlessness. Watching different bodies, we will see them obey very well Newton's first law. If at a certain moment of time a body is at rest with respect to the walls and ceiling, it continues to be at rest. Should we throw any material body, it will fly with a constant velocity until it collides with some other body or with the walls of the room, i.e. until it experiences a force of interaction with other material bodies. And it does not matter whether the room is at rest or is moving with constant velocity. In all these cases, Newton's law holds equally well, which confirms on the one hand Einstein's first postulate, and on the other hand the ability of our room to serve as an inertial frame of reference. Other natural phenomena in the room, including electric and magnetic phenomena, also satisfy the laws of nature quite successfully. One of the main characteristic features of the inertial frame of reference is the identity of all properties in all directions. We do not feel any difference whether it is the floor or the ceiling under our feet.

Now imagine that our room is a cabin of a space ship. Everything in it goes on as usually as long as the jet engine is switched off. But once the ship is powered, everything becomes different. We suddenly prove to be standing on the floor (the engine being just under the floor, the ship is accelerated in the "vertical" direction, i.e. from the floor to the ceiling) and we begin to feel a "force of gravity", which is the same as on the Earth, provided the acceleration of the ship equals $g = 9.8$ m/s. This force has nothing to do with Newton's law of gravitation, because it will vanish as soon as the engine is switched off. But nevertheless it acts exactly in the same way as a real force of gravity, and no device can detect any difference between what is taking place in our room now and what would have occurred in exactly the same room on the surface of the Earth.[1]

It is evident that, with the engine switched on, our room is not an inertial frame any longer. If we set free a certain fixed body, then, according to Newton's first law, it must continue to be at rest, but

---

1. In fact, though, there is a slight difference. The true gravitational force is directed to the center of the Earth. Therefore the gravitational forces, acting in the opposite corners of our room are not absolutely parallel as it would be on the space ship, but at a certain very small angle to each other. Later on we will give due attention to this "weak" effect. And until then we disregard it, though in reality it plays a very important role in nature. But it is hard to understand everything at once, and therefore we will refrain from considering it at this moment.

indeed, it will fall to the floor. A body thrown in an arbitrary direction will move not along a straight line, but along a parabola.

2.9.2. Meeting with free-fallers

While we were making our simple experiments, it became known that the space is inhabited. Our ship was visited by free-fallers — beings who live in empty space. Because the race of free-fallers has been conceived and evolved in the space, the state of weightlessness is quite natural to them and they cannot stand even the smallest force of gravity. Though we diminished the thrust of the engine, reducing our acceleration to $g/10$ (for us, it was almost absolute weightlessness), they were quite exhausted and asked us to switch off the engine altogether. They said they could normally exist only in inertial frames of reference and confirmed our conclusion that our room is an inertial frame of reference exclusively on condition that the engine is switched off.

During our talk with the free-fallers, an alarm signal was heard. Our stellar radar discovered the approach of a huge cosmic body — apparently an extinct star, possessing a very large mass. The computer informed us that, luckily, no collision was expected. The star would fly past us and, moving with a very high speed, it would not have time to entrap us into its tremendous gravitational field. But while it flew by us, our ship would be greatly accelerated, up to millions of g. First we were terribly frightened (no engine would be able to counteract such an acceleration), but the free-fallers calmed us down. They said that everything would be O.K. provided we did not switch on our ill-fated jet engine. When the star was flying by us, we would be falling to it freely with an acceleration as high as millions of g. But our fall being quite free, we would experience only weightlessness, i.e. we would not notice anything different from what experienced at the moment. If we did not look out of the porthole or look at the screen of the radar, we would never know that a huge body, causing such an enormous acceleration of our ship, was passing by us. The only possible danger might be expected at the end of the fall, which was fortunately out of the question. If we fell to the star, then, when touching it, our room would stop being an inertial frame. The collision would give rise to tremendous forces and accelerations that would be fatal not only to the free-fallers, but to us as well. The free-fallers assured us that the acceleration of a free fall, no matter how large it was, would be absolutely harmless to us, while the acceleration of a collision would be of mortal danger if it were even comparatively small, exceeding, say, just a few dozens of g. We wondered why the same acceleration in different cases would give so different results. They answered the question with a question: "With respect to what do you count your acceleration?" We had to answer that we counted it with respect to an inertial frame of reference.

"Just so, – the free-fallers answered gladly, – that's the thing! When we are falling freely, everything goes on just in the same way as if we were at rest or moving with constant velocity. Any frame when falling freely is always inertial and all accelerations must be counted with respect to it. And in a free fall the acceleration with respect to it is zero. That is why we feel fine. It would be quite different if, touching the surface of the star, we stopped abruptly. That would cause high accelerations which would result in irreversible consequences. We can say even more. Imagine you have miraculously survived the collision with the star. If even your landing were very soft, that would not save you. You and we are not the only bodies who want to fall freely. Every particle of the star wants to do the same. All the particles in a free fall try to occupy the same limited volume of space taken by the star. But the volume is limited, and after the particles have approached each other, there is no more room for them to fall any longer. There appear various forces – electromagnetic, nuclear etc. preventing them from any further free fall. Though the distances between the particles do not change any longer and they seem to be at rest, this impression is deceptive. Indeed, all of them experience very large accelerations by contrast to the case of a continuous free fall. Therefore even when stationed on the surface of the star, you will undergo so great acceleration that your body will fail to endure it."

We could not say anything against the plain and convincing arguments of the free-fallers, though the course of their reasoning sounded to us rather unusual. Even without their explanations, we knew a human being to be unable to stay at rest for a long time even on such a planet as the Jupiter where the gravitational force is only 3 times greater than that on the Earth. But we usually explained it not by the fact that the Jupiter accelerates us upward three times faster than the Earth (as we regarded ourselves as being at rest both on the

2.9. THE CURVED EMPTINESS

Jupiter and on the Earth), but by the fact that the Jupiter attracts us 3 times more strongly. When we displayed our point of view to the free-fallers, they asked us: "What is that gravitational force?" We explained them that it is just that force under the action of which a free fall is going on. The free-fallers couldn't help smiling, but did not want to tell us what it was that seemed so funny to them. They were afraid of offending us. At last we did persuade them to uncover their thoughts, and this is what they told us.

The most common game in their kindergartens is the so-called topsy-turvy. Due to weightlessness, the free-fallers experience great difficulties in orientation because, instead of four degrees of freedom in their motions, they have as many as six degrees. To ease their orientation, they make use of the Orienta which is the brightest star in their sky. Every free-faller, including even toddlers, wears a very compact device, that automatically orientates him in such a way that the Orienta is always beneath his feet. In the kindergarten young free-fallers make fun, switching off the apparatus and turning themselves "upside-down". Being young and inexperienced, a free-faller regards himself as the center of the universe and believes the whole world, rather than himself, to turn upside-down. Because in that position the world looks most unusual, this gives the children a lot of joy. When the Orienta proves to be over their head, rather than beneath their feet, this makes a lot of fun for them. The young free-fallers begin to worship the orientating device which, as they think, makes the world turn over. You just flip the switch, and the whole universe goes upside-down. They do not understand as yet that the orientating device has nothing to do with it, leaving alone that it is turned off.

While growing up, the free-fallers begin to realize their mistake and the game loses its charm. They then modify it in the following way. A young free-faller attaches to himself a small jet engine accelerating him up to $0.01g$ or so. While in the state of accelerated motion, the young free-faller thinks that it is not him, but the surrounding world which is accelerated. Of course, all bodies around him "accelerate" identically. The free-fallers become sure of it while watching their toys. All the toys begin "falling freely" with an "acceleration" of $0.01g$, irrespective of their mass. Children find it very funny and invent various reasons for that "acceleration". They say, for instance, that the engine, attached to them, generates an all-pervading gravitational field in the surrounding space and that that field acts on different bodies in full accordance with their masses, every force being equal to $0.01mg$. The greater the mass of the body, the greater the applied force, and at the same time the greater the inertia of the body. That's why all bodies "accelerate" quite identically. Until the free-fallers grow up, it does not occur to them that no forces act indeed on the surrounding bodies, which do not undergo any actual acceleration. All of them are either at rest or moving by inertia, whereas the true reason for what they see is the acceleration experienced by the observer together with his instruments and sense-organs, that serve him for perceiving the surrounding world. Such an explanation can be proved quite easily. Every free-faller can make sure of it even without special instruments, just due to his own super-sensitivity to any deviation from the state of weightlessness. It is sufficient to establish which of them is in the state of weightlessness – the observer or his environment. If the free-faller does not feel any violation of weightlessness, it is the objects surrounding him that are accelerated. In the opposite case it is she or he who is accelerated. It is impossible to distinguish the motion by inertia from the state of rest. But it is always possible to distinguish the accelerated motion from the motion by inertia even without going beyond the limits of the laboratory, given acceleration is caused by a non-gravitational force.

The game "topsy-turvy" has a lot of funny versions. Watching a body which is moving by inertia with a constant velocity, one can see that, relative to the observer, it moves along a parabola rather than along a straight line – just like a stone thrown horizontally on the Earth. Instead of a rectilinear acceleration, you may impart a slow rotation to yourself – round your own axis – and it will then seem to you that the surrounding bodies are orbiting round you just like satellites round the Earth. In all these situations, it is always possible to establish who is accelerating – you or your environment. Only the observer who is in the state of weightlessness can make a right estimation of the events that take place around him. And we, people of the Earth, would be able to have right judgments on everything, had we only ceased to be stationed in our earth laboratory and, together with that laboratory, started falling freely toward the center of the Earth. We would then see the stone, thrown horizontally, move indeed along a straight line and not along a parabola. And if, on the contrary, we are stationed on the surface of the Earth, we are actually accelerating upwards with the acceleration g relative to the state of a free fall, from which everything should be counted. It is due to that acceleration that we see everything in the wrong light. When a

2.9. THE CURVED EMPTINESS

body is at rest or is moving by inertia, we regard it as being accelerated and call its motion a free fall. As for the bodies that surround us in our laboratory, we declare them to be at rest, whereas in fact they are accelerating together with us. It is quite easy to make sure of it. Imagine a rocket on a terrestrial starting ground. The engine is switched on and the traction is equal to the weight of the rocket. The rocket is suspended in the air quite close to the surface of the Earth. It does not fly up or fall down, it just hovers there. If it were far from the Earth, it would certainly develop the acceleration g. But when close to the Earth, it is just suspended. However, inside the ship everything goes on just in the same way as outside. It turns out that when sitting on a chair in our room at home, we feel as if somewhere very deep in the bowels of the earth a jet engine is continuously working to prevent us from falling freely.

This kind of reasoning, displayed by the free-fallers, seemed most unusual to us. It turned out that a gravitational force did not exist at all and that people, like the children of the free-fallers, had just invented this extra physical quantity. Living on the Earth and trying to explain a free fall, they decided to make use of Newton's second law $F = mg$, where F is the gravitational force and g – the acceleration of a freely falling body. When doing so, they were in a non-inertial frame of reference, so that their instruments, being wrong, registered fictitious quantities F and g which in fact did not describe any reality. Had they themselves been falling freely together with their instruments, their frame of reference would be inertial. But their instruments would then have registered nothing but weightlessness: $F = 0$ and $g = 0$. Newton's law would hold, but it would look like this $0 = 0$. The two things that do not exist are equal to each other. But if it is true that we write down $F = mg$ only due to misunderstanding, being unaware of the fact that it is indeed we who are accelerated, while the freely falling bodies are just at rest, then, in a free fall, everything must accelerate quite identically. But it is just the way how everything happens, isn't it? Down, lead and water, when released in vacuum, accelerate identically, don't they?

But wait! Perhaps there is a straw here we can grasp at. If it is we and not the freely falling bodies that are actually accelerated, then even the objects that do not weigh anything must look accelerated, such as for example a ray of light, which is known to propagate rectilinearly, if watched by an observer who is at rest or is moving by inertia. If the ray of light was watched from a space ship, accelerated perpendicular to the ray, it would seem of course to propagate along a parabola. Can it be the same when we watch a horizontal ray of light while standing quietly on the surface of the Earth? Can it be that the ray of light, like a thrown stone, will also move along a curved trajectory? The free-fallers assured us that it is just the way how a real ray of light behaves and that we can easily prove it. It is enough to look at the part of the sky which is close to the Sun during a solar eclipse. We shall see the stars there, shifted from their proper places that they usually take according to the laws of celestial mechanics. It will be not the stars, of course, but their images that are really shifted. This shift will be caused by the fact that the rays of light, emitted by the stars and propagating past the Sun, will be falling freely toward the Sun and have their trajectories bent. Later on, we had a chance to prove experimentally this forecast. To be quite sure that light behaves in this way, we even climbed up a high tower and sent γ-rays from there directly downward. The crests of the γ-rays proved to fall down with the acceleration g. The rays were emitted by the nuclei of the atoms of radio-active iron, and the similar nuclei trapped the γ-rays below. The nuclei were tuned strictly to resonance, and even the slightest variation in the frequency of γ-rays during their free fall was enough for the nuclei-receivers to perceive γ-rays as aliens and stop responding to them. This variation was caused by the acceleration of the rays during their free fall from the tower, which made the crests of the rays at the base of the tower follow each other more often than near the top of the tower, so that they couldn't be accepted by the nuclei-receivers. To make these nuclei respond to the rays falling from the tower, they had to be acted upon in a special way. By the strength of that action it was possible to estimate the acceleration of the rays, which proved to be g.

We were also told that a substance, heated in the bowels of the Sun, looks a bit redder than the same substance, heated to the same temperature on the Earth. The rays of light, born on the Sun, can reach us only after they overcome the gravitation of the Sun. This makes them a bit slower and lowers their frequency, bringing it a bit closer to the red end of the spectrum. This is called a red shift of spectral lines. After that our opinion that light does not weigh anything has been radically changed. But the free-fallers argued against such a formulation. "How can something that is falling freely have any weight?" asked they. "Whatever is falling freely is always in the state of weightlessness, isn't it?" And we had to agree to that correction.

While we were having this very interesting and most instructive talk, we did not notice how the heavy star that had been threatening us passed by. Everything happened exactly in the way predicted by the free-fallers. We did not have any unpleasant sensations, though the instruments insisted that our acceleration in the field of the star was as high as million and a half of g. The instruments would also be unable to detect anything if not for their activities outside the ship. Engaged in the continuous location of the star, they recorded the change of the distance to it, which let them measure the value of our acceleration. We felt nothing because we were falling freely and thus belonged to an inertial frame of reference.

2.9.3. Returning to the Earth

Having thanked the free-fallers, we were about to start back. Some of them accepted our invitation and decided to visit the Earth with purely scientific intentions. They took with them some robots and a trove of sophisticated machinery, destined to overcome the force of gravity, or in terms of their language "for preventing a free fall from having its end". In other words, the free-fallers intended to disprove a well-known Russian saying: "Every fall has its end." While we were approaching the Earth, they sent their robots forward. On landing, the robots developed fabulous activities. In no time, they drilled the deepest shaft that transfixed the Earth. Then and there they built a cabin able to fall vertically down the shaft. On their landing, the free-fallers hurried into the cabin and indulged in their most pleasant occupation – a free fall. In their cabin, they passed through the whole globe. Falling through the center of the Earth, they first accelerated, then decelerated, and, having reached the opposite side of the Earth, they started falling down again. Interchanging their acceleration and deceleration, they were always in the state of weightlessness. Some deviations from that state, caused by the air drag, were compensated for by means of a small engine whose traction was controlled automatically. The further communication with the free-fallers was maintained by radio.

Their automatics for retaining weightlessness was very ingenious. A massive ball hung in the middle of the cabin. It was not fastened to the frame of the cabin and hovered on its own due to weightlessness.[1] A radar equipment mounted within the ball kept track of its distance to the floor and the ceiling of the cabin. Should the ball shift toward the ceiling or the floor, little engines were immediately switched on and either accelerated or decelerated the fall of the cabin. So the ball always hovered midway between the floor and the ceiling, and the free-fallers were blissfully happy in their state of weightlessness. It seemed unusual to us that the fall of the cabin was controlled by the ball located inside it. Nothing prevented the ball from falling freely, even the resistance of the ambient air, the air also falling together with the ball and the cabin.

While the free-fallers are falling in their cabin, let us sum up the knowledge and experience we have gotten during our space travel. They can be reduced to the following two rules:

1. The state of a free fall in a gravitational field is indistinguishable from the state of rest or motion at a constant velocity, taking place far away from great masses of substance.
2. The state of accelerated motion under the action of either electromagnetic or another non-gravitational force is indistinguishable from the state of rest in the relevant gravitational field.

A free fall is understood here in a wide sense of the word. It comprises any kind of the accelerated motion, taking place freely in the gravitational field, such as the motion of the thrown stone, the orbiting of a satellite round the Earth, of planets round the Sun, etc. A total absence of any forces but gravitational is the only condition that must be satisfied to make the fall free. The rotation or the oscillation of a load, suspended on a rope, must be excluded, because there are electromagnetic forces in a stretched rope, that change the direction of the velocity of the load. The rotation of a flywheel cannot be incorporated either, for the same reason. Both these examples obey not the first, but the second rule given above.

From these two rules, taken as they are, it follows that gravitational forces do not exist in nature. Indeed, in a free fall, they do not manifest themselves (everything goes on as if it were a motion by inertia with a constant velocity), and in the case of the body sta-

---

1. This example was borrowed from "Spacetime Physics" by E.F.Taylor and J.A.Wheeler, 1966.

2.9. THE CURVED EMPTINESS

tioned in a gravitational field, they are also unnecessary (everything can be explained without them if the explanation is given in an inertial frame of reference, i.e. in a frame being in a free fall). It turns out that gravitational forces are fictitious as they arise only in non-inertial frames of reference. Very long ago, before Newton formulated his laws, it had been believed that even motion with a constant velocity is always caused by the action of some force. If there is any motion, there must be a force that is responsible for it. Newton's great predecessor Iohannes Kepler, who was the first to formulate the laws of the revolution of planets round the Sun and to develop the theory of solar eclipses, declared that in celestial mechanics these forces were contributed by angels. Newton rejected such ideas and preferred the concept of inertia as a reason for a free motion with constant velocity, which was a crucial step on the way to the classical mechanics. Einstein went still further in this direction. He declared that not only a motion with constant velocity, but also a free fall goes by inertia, i.e. without any external forces. This step proved to be crucial for the development of the new, Einsteinian theory of gravitation. After the development of special relativity all the old laws of the classical physics remained safe and sound, though many terms got a new meaning there. They remained alive even after the development of Einsteinian theory of gravitation – all of them, but one. Newton's law of universal gravitation had to be revised in the most radical way. It was deprived of its main component – the gravitational force, that disappeared altogether. Now we can explain why, when studying relativistic effects, we did our best to avoid gravitational forces. Had those forces been used, they would have to behave like electromagnetic forces every time when the body changed the uniform velocity of its motion. The longitudinal forces would have to be retained, and the transverse forces would have to relax by a factor of γ. Otherwise the principle of relativity would have been violated. Electromagnetic forces have their own good reasons to behave just in that particular way. Gravitational forces do not have such reasons unless, of course, relativity itself is regarded as such a reason. It is clear now why these reasons have not been found: There are no forces there, so that any question about the forces becomes absurd. As for the pendulum clock which we had to place on board the accelerating space ship when we studied the frequency of its ticking in Section 2.3, it can now be safely transferred to the Earth. According to the second rule nothing new will happen to it.

So, there are no gravitational forces in nature. Does it mean that gravitation itself does not exist either? Had we accepted the two above-given rules without any restrictions, then there would be no place for gravitation in nature. All the phenomena could be explained very well without it. A free fall would have been reduced to the motion with constant velocity, and the bodies on the Earth would have weight not because they are attracted by the Earth, but because all of us are accelerating in comparison with the case of a free fall, that should be regarded as standard. Everything would be just so if not for one "small" circumstance that was mentioned above in the footnote on page 210. Now you will see how this minute restriction will grow as large as the universe.

2.9.4. A mysterious tickle

To better understand the essence of the issue, let us assume that the world is quite different from what it is. Suppose the universe consists of nothing else but the Earth and the space. Let the Earth have the shape not of a sphere, but of a thick disc, whose diameter is infinitely large. The thickness of the disc should be chosen in such a way that it will correspond to the acceleration g of the free fall on the surface of the Earth. In such system all bodies will fall freely along straight lines that will not intersect in the center of the Earth, but will be always parallel to each other and perpendicular to the plane surface of the Earth. Hence, the acceleration of a free fall g will not depend on the distance to the Earth and will play the role of a world constant. Let us make it clear from the beginning that, in such universe, the Moon would not be able to move round the Earth and would be sure to fall on it, moving in a parabola. If the world were such, the rules formulated above would be valid without any restrictions, leaving no room to gravitational forces as well as to gravitation itself. In that world a free fall would be indistinguishable from the motion by inertia with constant velocity. If the observer belonged to the inertial, i.e. freely falling frame of reference, he would then see every body move with a uniform velocity until some external forces are applied to it (for example, until it touches the Earth). If a stone were thrown horizontally in such system, then any observer falling freely, no matter how far from the Earth he were, would see the stone move not in a parabola, but in a straight line.

Let us now return to our real world and ask ourselves: "Can a freely falling observer in our world see the stone move not rectilinearly but in a parabola?" If the observer is falling near the place where the stone is moving, he will certainly say that the stone is moving along a straight line, i.e. according to him the motion of the stone is rectilinear and takes place at a uniform velocity. But the farther the observer is removed upward from that place where the stone is moving, the smaller is the acceleration g_1 of his free fall, and so he will soon be able to notice that the trajectory of the stone is not a straight line, but a parabola, whose curvature is determined by the difference $g - g_1$ between the acceleration of the stone and that of the observer. Now the observer sees that the velocity of the stone is not uniform. He can measure the acceleration and it will be equal to g, provided the acceleration g_1 of the observer himself is negligible. We must believe him because he is in an inertial frame of reference. The stone moves by inertia just as before, though this time the observer may wonder what it is that accelerates the stone and curves its trajectory when there are no forces acting on it. Should we introduce again the gravitational force as the cause of the acceleration of the stone? But the observer who is falling freely close to the stone might object to it. For him the presence of that force would seem unnatural. For he would then have to admit that the stone is moving by inertia in spite of the force applied to it.

It was not at once that Einstein found a way out of this puzzle. It seemed to be a deadlock: there is no force but there is gravitation! This gravitation can be not only seen from afar, but also felt directly, with shut eyes, even in the case of a fall which is expected to be quite free. To make sure of it, it is enough to return to our free-fallers who continue their falling freely and to ask them whether they feel anything unusual besides the blissful sensation of weightlessness. The radio operator, upholding the communication, informed us that something was wrong there. He heard the roars of hysteric laughter from the cabin. The intensity of the laughter was proportional to the speed of fall of the free-fallers. When the cabin flies upward, lowering its speed, the laughter gradually weakens and even vanishes. But as soon as the cabin begins to speed down, the laughter is resumed and intensified proportionally to the velocity of the cabin. Having caught the moment when they stopped laughing, we asked the free-fallers to tell us the reason for their strange behavior. It proved to be some unusual tickle that all of them experienced without understanding the true source of it. Something tickled them very deeply from inside. Some of them said it arose in the breast, others — in the throat, somebody else said it went from the stomach. Their further explanations were drowned in gusts of laughter, for at that moment the cabin sped down.

Soon we managed to get at the reason. It was of an earthly character, so we could understand it better than the free-fallers. Suppose we have lifted two stones at a height of 5 kilometers above the surface of the Earth, the stones being horizontally 1 kilometer apart. If the two stones are dropped, then during their vertical fall, the horizontal distance between them will become about 1 m shorter. That is caused by the fact that the Earth is a globe, and the stones fall along the straight lines converging to the center of the Earth. The particles of the bodies of the free-fallers also tend to fall along the converging trajectories. The inner organs trying to come closer to each other in a horizontal direction, it gives a sensation resembling a tickle. It seems most strange and queer to the free-fallers, for they have never experienced it in the space. A free hovering in the space, and a free fall on the Earth proved after all two different things. Though the difference was not great and it was the free-fallers with their high sensitivity that could spot it.

It's just a proper time now to confess that we also experienced the same tickle when we were falling freely in the space to the huge star. We were then prudently silent about it. We could not discuss everything at a time. A relevant warning was made in footnote on page 210.

Our old Mother-Earth in its revolution round the Sun also experiences such a kind of tickle. At the first sight it seems that the solar gravitation should not be felt at all on the Earth. For, in Newtonian terms, the centrifugal force cancels the force of solar attraction or, in terms of Einstein's approach, there is no force of gravitation. But these statements are exactly valid only at a certain point of the Earth which is somewhere inside the planet. On the surface of the Earth these conditions are somewhat violated. On the day-side of the Earth the gravitation of the Sun is a bit stronger, and the centrifugal force — a bit weaker than at the center of the Earth. On the night-side of the Earth the same effect takes place, though the other way round. Due to this, different parts of the Earth tend to

2.9. THE CURVED EMPTINESS 223

shift relative to each other, which gives rise to high and low tides as well as to shifts of strata in the earth crust.[1]

The smaller is the size of a planet, the weaker are these effects, provided other presumptions are the same. The "pure" weightlessness may be observed only in the case of sufficiently small bodies. It may be realized only in a sufficiently small volume. If, when falling freely, one does not go beyond the boundary of a small region of space, then it may seem to him that there is no gravitation. But should he extend his excursions far beyond this boundary, the gravitation would manifest itself by means of tides, by transverse squeeze of a freely falling body and by other similar phenomena. While watching the thrown stone, we observed the same regularity. While we were falling close to the stone, we did not notice the curvature of its trajectory. But as soon as we flew well off into the space, we began to notice it from there quite distinctly. We saw that the stone flew not in a straight line, but in a parabola. If we kept on explaining it by the force of gravity, we would have to admit that the further from the stone the observation post is located, the more noticeable the force becomes. When ascribed to a force, such property looks very strange.

In addition to this, it is important to note that when the observer goes away from the Earth, he begins to notice the curvature of the trajectory of the stone not at once, but gradually. Even when close to the stone, he can use very sensitive instruments and thus learn that the trajectory of the stone is slightly curved. He can discover it due to the fact that the acceleration of the free fall of the stone and that of the observer are slightly different from each other. Therefore it can be said that in different regions of space the acceleration of a free fall is different, and this difference is detectable. But this can be detected only from a neighboring region of space, remote enough for the measuring instruments to register it. The farther from the stone, the greater the difference, until at last the motion of the stone is seen on its "true" scale, i.e. as a parabola whose curvature is determined by the acceleration g. This net result can be gotten only after the observer is beyond the region, "filled" with gravitation, hovering in the "pure" space, far away from gravitating masses. If he is removed

---

1. It is the Moon that makes the main contribution to the ocean tides, but we will leave that aside so as not to be distracted from the essence of the issue.

still further, the trajectory of the stone will not experience any further change, unless he happens to get under the action of some foreign gravitating object.

2.9.5. Curved spacetime

Let us now take a look at the gravitation from another side. Suppose somewhere in the cosmos, there occurred fast shifts and displacements of some huge masses. Can we learn of them instantly? If yes, then a signal propagating faster than light is in our hands, and the principle of relativity must be revised. The previous sections of the book tell us that such supposition is hardly credible. It would be much more reasonable to suppose that, like in the case of interacting electric charges, the information about the location of gravitating masses propagates with a certain finite speed c_g either smaller than the speed of light, or equal to it. But the direct cause of the gravitational action upon a certain body should be sought then in the vicinity of that body. Perhaps the Earth generates its gravitational field in the surrounding space, and the free-fallers sense it in the form of tickling, even when they are falling freely. But if it is a field of something, then what is this "something"? An electric field, for example, is a field of forces: At every point of space there is a certain vector force **E**, acting on a unit stationary test charge. In the case of gravitation, there are no forces – we have already agreed about it. But if not forces, what is it then?

Einstein reasoned like this: What we are in search of must, first, be associated with a small region of space in the vicinity of the body, acted upon by gravitation. Secondly, it must be undetectable, or hardly detectable if the observations are localized in this small region of space. Thirdly, it must become always more noticeable as the region of the observation is widened. The word "space" is involved in all the three requirements. This suggests that we are in search of some property of space itself. And Einstein has discovered at last this property. It was the curvature of space, which long ago had been discussed by mathematicians without any connection with the problems of gravitation.

2.9. THE CURVED EMPTINESS

To understand this idea we will have to simplify our world to a great extent. We will reduce it to a sheet of paper, and ourselves – to flat insects, crawling about this sheet, incapable of looking at it from above. As for Einstein, let him stay in the three-dimensional world and observe our actions from there.[1] Crawling about the sheet of paper, we can examine it. We will discover that it has edges and that it has the shape of a rectangle. We will be able to draw various plane figures on the sheet, developing plane geometry. We will be able to introduce a concept of a straight line, defining it, for example, as the shortest way between two points. We will be able to measure the sum of the angles of a triangle and make sure that it is equal to 180^0. We lack imagination to visualize space figures, such as a sphere or a cone, but we will be able to formally build a logically consistent space geometry and derive the formula, say, for the volume of a sphere, though we cannot imagine what it looks like. We can only state that it is a three-dimensional analogue of a plane circle, which is quite familiar to us.

Among our geometrical achievements there will also be a concept of parallel straight lines. We will give them a definition. We will say, for example, that two straight lines are parallel provided there is a third line, perpendicular to both of them. We will learn from practice that two parallel lines never intersect. We will make sure of it in the following way: Two of us will take some initial positions in two different points. We will stretch a string between us, or will send a ray of light. Both of us will then begin moving with a uniform velocity in the direction perpendicular to the string, connecting us. Moving in this way, we will see that the distance between us will all the time be the same, so that our ways will never intersect.

Now imagine that Einstein, wishing to check visually the correctness of his suppositions, suddenly curved the sheet of paper we are on. Suppose he gave this sheet the shape of a sphere, as the simplest version of a curved volumetric figure. Shall we learn that our world has been curved? Crawling about a small area, we will learn nothing. If the size of the area is small in comparison with the radius of the sphere, we will still regard our world as being plane. The precision of our instruments will be insufficient to detect the curvature. In a

---

1. The rest part of this subsection is a paraphrase of the explanations presented in the books of E.F.Taylor and J.A.Wheeler. These explanations are so perfect that it is hardly possible to invent a better way of envisioning these effects. See, for example, "Spacetime Physics", 1966 by E.F.Taylor and J.A.Wheeler .

small region the geometry will be the same as on a plane sheet of paper. The sum of the angles of a triangle will still be 180^0. And moving parallel to each other, we will see that the distance between us does not change. But that will be so as long as the region of observation is small. Prolonging the routes of our travels and plotting, on our sphere, a triangle whose sides are comparable with the radius of the sphere, we will discover that the sum of the angles of the triangle exceeds 180^0 by quite a noticeable value. One can easily make sure of it, taking an isosceles triangle whose basis lies on the equator of the sphere, and the apex is on the pole. The sum of the two angles adjacent to the basis of the triangle, taken alone, is already equal to 180^0 without even taking into account the third angle whose apex is fixed at the pole. If the length of the basis of such triangle is equal to a quarter of the length of the equator, then the angle at the pole will be 90^0 too, and the sum of the three angles will be equal to 270^0. Increasing the length of the basis to almost the full length of the equator, we will have the angle at the pole equal to almost 360^0, and the sum of the three angles in the triangle will be almost 540^0. There will be also other surprises in store for us. For example, the ratio of the length of a circumference to its diameter will not be equal to π any longer. We have to revise the whole geometry.

If we were not insects, then, having risen to our full height, we would be able to survey the line of the horizon and see it widen when our head is lifted above the surface of the sphere. That would be the clue for understanding all the tricks mentioned above. But according to the assumed agreement, we are flat insects, unable to raise our head even a little bit, and therefore we wonder what has happened to our world. We would wonder even more if we made a tour round the world. Moving all the time along a straight line, we would find ourselves at that very point from which we started. We would not be able to discover any borders of our world.

Let us see now what has happened to the parallel lines. Two of us, my friend and I, take our initial positions at two spatially separated points on the equator stretching a string between us. Then we begin our motion with a constant velocity v, perpendicular to the string. We should take care of our routes to be always perpendicular to the string. It is the same as to go always North. Evidently, our routes will be coincident with the relevant meridians. We will think that we move along straight lines, not noticing the curvature of our path. We know that a straight line is the shortest path between the

2.9. THE CURVED EMPTINESS

two points, and on the sphere the shortest path always coincides with the principal circle, i.e. the circle whose radius is equal to the radius of the equator. When a plane starts from Moscow to Vladivostok, it does not fly East, it flies North-East or so, taking to the region of rather high latitudes. The navigator knows that a straight line, extending from Moscow to Vladivostok on a flat map, is far from the shortest way between them because the flat map does not account for the curvature of the surface of the Earth. To reach Vladivostok in the shortest time, the pilot must keep to the principal circle. Through any two points on the surface of a sphere it is possible to draw only one such circle. In the case involved, the role of the principal circles will be played by the relevant meridians. Moving along them, we will soon see the distance between us gradually reduce, in spite of the fact that our routes seem to be straight lines and are always perpendicular to the straight line connecting us. If we do not get too far in the region of high latitudes, the distance S between us will then change according to the following approximate law derived in accordance with the rules of trigonometry

$$S \cong S_0 - \frac{S_0 v^2 t^2}{2R^2}. \qquad (2.54)$$

Here S_0 is the initial distance between us, t is the current time and R – the radius of the sphere, about which we, being flat insects, haven't got even the slightest idea. Perhaps formula (2.54) is familiar to you. It is a length-time dependence for a body moving with constant acceleration. The role of acceleration is played here by the value

$$g = -\frac{S_0 v^2}{R^2}.$$

We will look for the reason for the acceleration. Not seeing the true reason, we might say, for example, that there is a force of mutual attraction acting between us. Because the distance S contracts not only between us, but also between all the other bodies that accompany us, we will have to declare that the magnitude of the attracting force is proportional to the masses of the bodies: $F = mg$. That's why all the bodies "fall" toward each other quite identically. If only we could have a look out of our world onto the three-dimensional world, we would see Einstein, smiling there when listening to our reasoning.

But we are not inclined to joke. We feel some defectiveness in our reasoning. We do not understand, for example, why the force of attraction of two bodies F is proportional to the squared constant speed v of their motion along the two meridians. And not only that. Prolonging our travel into the region of high latitudes, we discover some deviations from the law (2.54). But we cannot get at the reason of that deviation. And here mathematicians come to our aid. (In our three-dimensional world it was successfully done by Rieman and Lobachevsky) They say: "We understand that our real world is two-dimensional. This is a matter of fact. But it was long ago that mathematics in its abstract formalisms went well beyond the limits of that two-dimensional space. It succeeded in describing various three-dimensional figures and tells us how to find their shape and volume. A sphere is the simplest among them. We cannot imagine what this sphere looks like. Because there is nothing like that in our world. But we can describe this sphere in purely mathematical terms. And if we assume that our two-dimensional world is the surface of such a sphere, then the conformity of (2.54) with the law of motion can be explained only geometrically, without any attracting forces, which indeed, do not exist at all and have been just invented by us. It is very easy to check up our assumption. It is enough to make a round-trip about the world. And if, moving all the time forward along a straight line, we come once again to the point of our departure, there will be no doubt about it. Our world is curved. And though we cannot imagine this curvature visually, we come across it at every step."

Our real world is three-dimensional. But who can guarantee that it is not curved? It is hard to imagine what this curvature would look like in a four- or five-dimensional space, though we come across this curvature daily. Einstein was the first to notice it. Should we travel about the universe, moving always along a straight line, perhaps we might make a round-trip there. Though the path would be too long. Mind that the length of the observable part of the universe is about 10 billion years taken by light to cover that enormous distance. And there is no guarantee that our travel will be indeed a round-trip, for the Universe may prove to be curved but not closed. In our simplified story the sphere was taken just as an example. The sheet of paper could have been curved in quite a different way. And if our universe is yet closed, there being many arguments in favor of

2.9. THE CURVED EMPTINESS

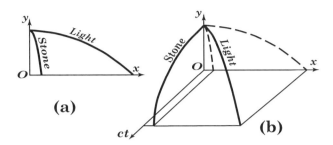

Fig.31: **(a)** The trajectories of light and stone ejected horizontally in a gravitational field directed vertically. **(b)** The same trajectories represented in a three-dimensional chart with reduced time ct as a third coordinate In contrast to the two-dimensional representation (a), the trajectories of light and stone in representation (b) turn out of the same size, which suggests that gravitation propagates with the same speed as light.

it, then it may be that, looking through the telescope at some distant galaxy, we are looking at ourselves. But we cannot see ourselves, because the distance is too long.

But if gravity is explained by the curvature, then, under its action, the trajectories of different bodies at a certain point of space must be curved identically. Let us see if it is the case. Let us compare the trajectory of a stone thrown horizontally and that of a ray of light, emitted in the same direction from the same point.They are shown in Fig.31(a). Both of them are parabolic. But how different their curvatures are! Unlike the stone, light propagates along a line, almost straight. The difference is so great, that for clarity Fig.31(a) had to be given not to scale. Too great is the difference between the speed of the stone and that of light. Does it mean that gravitation cannot be explained by the curvature of space?

Let us consider more deeply what it is that is shown in Fig.31(a). Is it a true picture of motion? What we see is in fact not a vivid motion but just a dead trace of it. If that flat picture were drawn by an artist rather than a scientist, it would be performed in quite another way. Using some special artistic technique, the author would have emphasized that the propagation of light is many times faster than the flight of a stone. He would not forget the role of time, which has been missed by us. Light covers a certain distance much sooner than the stone. And this is not reflected in Fig.31(a).

To rectify this slip, it is sufficient to introduce time into Fig.31(a). In addition to the two coordinates x and y, it is necessary to introduce the third coordinate — that of time t or, still better, ct, for all the coordinates to be presented in the same units. Now the trajectories of the stone and of the ray of light will be displayed not on the plane xy, but in the space of the coordinates x, y, ct, as shown in Fig.31(b). At either point of the trajectory of the stone, this time tells us not only where the stone was, but also when it was there. As a result of it, the shapes of the trajectories have apparently changed. The trajectory of the ray of light, in comparison with Fig.31(a), has changed its curvature but little, while the trajectory of the stone has become much more straight than it was in Fig.31(a). Now its curvature approached that of the light. This suggests that the speed of light c is a fundamental constant not only in electrodynamics, but also in gravitation, which on the face of it seems to have nothing to do with the propagation of light. That is why we, even in advance, are so sure that gravitational disturbances, whose propagation has not yet been registered experimentally, are also transmitted through emptiness with the speed $c_g = c$, equal to the speed of light.

So, the curvature of space does indeed explain the gravitation. But this space cannot be separated from time. We must speak of the curvature of spacetime, which is three-dimensional in Fig.31(b), and four-dimensional in a general case. Mathematically, this curvature is described in a rather sophisticated way. The curvature of a line or a surface is characterized in mathematics by one number, while the curvature of a three-dimensional space — by six numbers, and the curvature of a four-dimensional space — by twenty numbers. But all those complications are of a purely formal character and, from the standpoint of physics, do not add anything essentially new to what has been said above. Much more important is another thing. The curvature of the world space does not arise by itself. It arises under the action of large masses of the substance, distributed about the universe. There appears a closed circle in our reasoning: Bodies, being displaced in the space, affect the curvature of spacetime, and the curvature, in its turn, affects the motion and consequently the disposition of the bodies. If there are many bodies, participating in the gravitational interaction, it makes the solution of the problem rather difficult. The situation becomes still more challenging when we try to find the solution for the whole universe. It has not been possible so far to get an unambiguous solution. Even if Einstein's equations are successfully solved, it does not bring us to a single result. Our knowledge of the universe is too poor as yet, and we have not sufficient information of what we have to substitute into the equations. And yet, Einstein's theory succeeded in predicting and explaining many things that did not follow at all from Newton's law of gravitation. One of the possible solutions of the gravitational equations results in the continuous expansion of the universe. Astrophysical observations of the spectra of distant galaxies confirm this result. Galaxies are indeed expanding in all directions, and the more distant they are, the greater is the speed of their expansion. As for our Solar system, Newton's law is valid there with a high accuracy, and Einstein's theory introduces but very small corrections. For the Mercury those corrections proved to be observable and had been registered by astronomers a long time before.

2.9.6. Summary

The theory of gravitation in the form Einstein has given it, is not yet a finished chapter of modern physics. Therefore we will highlight here but its most simple and indubitable theses:

1. *In a sufficiently small region of space, the state of a free fall of bodies or systems under the action of gravitating masses is indistinguishable from the state of rest or motion by inertia beyond the limits of action of those masses.*

2. *In a sufficiently small region of space, the state of rest of bodies or systems within the limits of action of gravitating masses is indistinguishable from the state of accelerated motion beyond the limits of action of those masses.*

3. *The spacetime is curved by gravitating masses, which brings about two consequences:*
   a) A free fall can be distinguished from the rectilinear motion by inertia, if observed from neighboring regions of space.
   b) The free fall of bodies of large dimensions gives rise to tide phenomena inside these bodies.

The curvature of space results also in many other consequences of great importance for the behavior and properties of the universe, but we will not dwell on them here.

Conclusion

Relativity dethroned the absoluteness ascribed earlier to a lot of physical quantities and objects. Such firm pillars of classical physics as absolute time, absolute space, and the world ether collapsed. Lorentz's ether proved to be undetectable, while time and space, if considered separately from each other, proved to be dependent on what frame of reference they were observed from. Even such terms as *now* and *then, earlier* and *later*, which had seemed obvious, had to be reconsidered, if assigned to different regions of space. It turned out that, being applied to the same events, all of them could be different, depending on the frame of reference they were considered in. The same fate befell other physical quantities, such as mass, energy, momentum, force (that in the case of gravitation disappeared altogether) etc. Does it mean that relativization (i.e. making absolute concepts relative), started by Einstein so successfully, should be continued by his followers until the last absolute concept disappears for long? Won't such a victory mean the complete and ultimate triumph of science over prejudices? If it does, we must promptly take on those absolute quantities that "occasionally" have happened to survive (for example the net electric charge), and do our best to make them relative by any means.

Such an approach to relativity would be absolutely wrong. Relativity does not abolish the concept of absoluteness, It only introduces new absolute quantities instead of the old ones. In pre-einsteinian physics, space and time were believed to be absolute. Now they are not. Now it is their combination − the interval − that proves to be absolute. (We tried to describe it in a simplified form in Section 2.7). In the past, energy and momentum were regarded as being absolute. Now it is their combination that is absolute − the so-called energy-momentum tensor − we had no chance to discuss it here (the same refers to many other invariants). It is too difficult to discuss it without a relevant mathematical apparatus. So we will just mention here that such absolute mathematical representations exist, that they do not depend on the frame of reference, and that most of them have been born by that very theory whose brand name, given by Einstein, is Relativity. It was not by chance that Einstein has chosen that title. His main task was to demonstrate and well-ground the relative character of rest and of free motion. This task was not only fulfilled, but even overfulfilled by him. He has shown that all the phenomena in nature proceed in the same way not only in the inertial frames of reference that are moving with constant velocities, but also in those which fall freely in a gravitational field, provided the size of the region where these phenomena take place is not too large. It was also shown that a state of rest in the zone of gravitation is equivalent to the accelerated motion outside that zone. This equivalence was a starting point for the discovery of the curvature of spacetime as being the essence of gravitation. But that proved to be the end of relativization. Nature firmly stood up to further advance in that direction.

The fact that a free fall is indistinguishable from the motion with constant velocity was established only for a sufficiently small region of space. But even in a small region the triumph of relativity was not quite complete. On the one hand, we could state that the size of the body involved could be made so small that tidal phenomena would not be felt, no matter how sensitive the measuring instruments might be. On the other hand, the inverse statement was obvious too – however small the size of the body involved might be, the sensitivity of the measuring instruments could be generally made high enough to detect the tidal phenomena. This idea can be also formulated in another way: Widening the region of observation (or the size of the body), it is always possible to learn which region of space is curved, and which is not. Therefore, some scientists refrain from applying the term "general relativity" to Einstein's theory of gravitation. They regard the term "relativity" as being applicable only to the case of motion at a constant velocity, using this term without the adjective "special", that becomes unnecessary. Such terminology emphasizes the idea that there is no other relativity but special relativity – all the rest refers to Einstein's theory of gravitation.

Absoluteness manifests itself in nature not only through abstract mathematical formulations, existing in our imagination or on a sheet of paper. It is felt in an immense number of various natural phenomena, though it is not easy to establish its roots and limits as certainly and unambiguously as we would like. Let us turn, for example, to a pendulum, swaying on the North pole. During a revolution of the Earth, the plane of the pendulum holds its position relative to the stars, and therefore is turning relative to the Earth making one revolution a day. What physical object is felt by the pendulum, when it retains its plane in space unchanged? Relative to what does it remain unchanged? The massive gyro, successfully used in navigational instruments, behaves similarly. If such a gyro is placed inside a sealed hull, and is floating indifferently within a liquid, then the axis of the gyro always retains its direction relative to the stars, irrespective of the fact how or in what direction the aircraft, ship, rocket or spacecraft with this gyro has turned. What is it that prevents the axis of the gyro from turning together with the aircraft? You might remember the design of the automatic device of the free-fallers (see Section 2.9) that was upholding the free fall of the cabin despite the air drag in the shaft. At the middle of the cabin hovered a massive ball, which was not fixed to anything. It was that ball that served as a standard, supporting a free fall of the cabin and providing weightlessness inside it. And what supported the ball? Whatever it might be, it was not the cabin.

To feel the roots of this problem, imagine a droplet of liquid, placed somewhere in the space far away from gravitating masses. Under the action of the surface tension such a droplet will take the shape of sphere. If the droplet is made to rotate round its axis, then it will flatten at the poles and turn into ellipsoid under the action of centrifugal forces. The faster the rotation, the stronger the flattening. Among all possible speeds of rotation there is one and only one speed that permits the droplet to retain its spherical shape. This speed is declared to be equal to zero and is used as a standard base which all the other speeds of rotation are referred to. This is that very absolute reference which we lack in the case of uniform velocity. In the case of a rotating body, it exists. But if it does exist, what physical object is it associated with? Can it be that it is Lorentz's ether that manifests itself? Why then doesn't it manifest itself in the case of uniform velocity? Newton thought that the droplet feels the absolute space, though he could not explain how that absolute space is arranged. An opposite point of view was set forth by Mach at the end of 19th century. His idea, for a certain time, was shared by Einstein who did a lot to ground it. (Unfortunately, his results were too far from what he had expected). Therefore, it is sometimes called the Mach-Einstein principle. According to that principle the drop feels

CONCLUSION

the whole universe. It flattens only on condition that it rotates relative to the whole universe. In scientific wording, the term "universe" is often replaced by "distant stars" (as was customary to say in old times), or "distant galaxies" (as is said nowadays). But whatever it is called, it refers to the whole universe. If the whole Universe had contained nothing but our droplet, then, according to that principle, the droplet would not be deformed at all by its rotation. According to this principle the inertia of bodies exists and manifests only as long as there are remote huge gravitating masses. The distant curved regions of the world space somehow extend their influence even to those regions that are beyond their gravitation. The Mach-Einstein principle can be verified experimentally. To do it, it is enough to remove all the galaxies to infinity. You might be smiling. But the universe is expanding, and every year the galaxies withdraw from us farther and farther. If the Mach-Einstein principle is valid, their influence must get gradually weaker. For instance, the Earth must every year retard its revolution by about 10^{-10}, and the distance between the Earth and the Moon must decrease several centimeters a year. The measuring instruments available nowadays are almost ready to register such negligible values. The main difficulty is not in the measuring instruments but in the necessity to separate the desired result from the influence of the tides, that make a commensurable contribution into the secular variations of terrestrial and lunar daily cycles. But in due time the influence of tides may be clarified, paving way for the most crucial measurements in the fundamental physics. The progress in the methods of measuring and in the relevant technology gives hope that this will happen in the sensible future. Scientists are not unanimous in predicting the result. It is quite possible that the droplet would flatten even if it were the only material object in the whole universe. It does not contradict Einstein's theory of gravitation. Such behavior of the droplet would mean that when revolving, the droplet feels not the whole universe, but the world space in the vicinity of the droplet – that very space whose curvature tickles the Earth with high and low tides. If something turns out curved, then this something has to exist, doesn't it?

We deliberately finish this book with questions that have not yet gotten an unambiguous answer. We want you to realize how boundless is the ocean of the Unknown, surrounding the cozy and cultivated islet whose name is "Special Relativity".

Birth and evolution of non-postulated relativity

A historical review

1. How classical physics and special relativity found themselves in opposition to each other

The search for the good reasons underlying the relativistic effects, actually started long ago – at the time when Galileo's principle of relativity failed to withstand the pressure of electrodynamics and Einstein's relativity had yet to be created [1],[2],[3]. It was then that both the Lorentz length contraction and the mass-velocity dependence were discovered in electrodynamics. It was too bad that no one dared then to extend the mass-velocity dependence to neutral bodies, which were already known to contain a lot of electricity and hence internal magnetic fields. That would inevitably lead to what is now called "relativistic dynamics"[1] with the ether still remaining as a background of what is called now "relativistic effects". However the blind eye could be turned to that ether, which would all the same fade away as soon as "someone" discovered the remarkable symmetry of the properties of the instruments, moving through the ether. That symmetry would lead that "someone" to the idea that there would be no difference in the results of the measurements between the two opposite occurrences: the first – with the instruments fixed to the ether and the bodies observed moving through it, and the second –

---

1. Studying the textbooks, one has a chance to be taught that special relativity changed Newtonian mechanics. Actually, the equations of motion have not suffered any change. It was only their solutions that really changed as soon as the mass-velocity dependence was taken into account. As for the equations of motion, Newton formulated them in a way which permitted the mass to undergo any variations. It was only Newton's law of gravitation that was given up, but this happened out of the scope of special relativity.

with the instruments and the bodies interchanged. That symmetry would not disprove the existence of the ether, which would become only useless but not harmful. It would be only a spare pillar bearing no real burden and sometimes being even helpful in visualizing a relevant reasoning. (Isn't it easier to agree with the reality of the length contraction or clock retardation, when they occur under the action of the ether drift with the ether fixed to any inertial frame of reference one likes? If yes, then one may make use of the ether – it will not influence the results in any way. If not, then one can do without any ether, making absolutely the same derivations and finally arriving at exactly the same result.)

If it were not for Einstein and his postulates, this work would have run on and on, and the time dilation would have been discovered with the ether used as a background for the relevant derivations. This time dilation would include the dependence of the clock reading on the position of the clock on the platform which is speeding through the ether. It had been called a local time by Lorentz, though he had never ascribed a real physical meaning to it [2]. Due to relativity, we can now be certain that the identical dependence would have been discovered for the clocks fixed to the ether and observed by the instruments quickly moving through it, so that the ether would inevitably have faded away [4]. The time-space dependence would have been called The Relativity of Simultaneity, and Einstein's postulates would have entered science just in the same way as classical physics had been enriched long ago with the principle of energy conservation and all the other conservation rules. The special theory of relativity would have been created and would most probably not have been called a theory, but rather a principle – the Principle of Relativity – just the term proposed by Einstein himself in his pioneer work [5] and only later replaced by the term "theory" mostly in connection with general relativity.[1] It would be then that Newton's law of gravitation would be given up as incompatible with the principle of relativity as well as with Newton's equations of motion and Maxwell-Lorents's electrodynamics.

---

1. Though Einstein often applied the term theory to special relativity in order to avoid its separation from general relativity, he always regarded his postulates as a restricting principle rather than the fundamental equations which predetermine the properties of nature. He wrote about it explicitly in his "Autobiographical Notes". A detailed consideration of it will be given in Subsection 2.

2. WHAT EINSTEIN THOUGHT ABOUT IT

But history went by another way. So revolutionary and so effective were Einstein's postulates, that the classical interpretation of the relativistic effects was given up together with the ether. The success of Lorentz in the classical interpretation of the length contraction and mass-velocity dependence looked then as only an accidental particular achievement to be honored and shelved. This undeserved fate befell not only the mass-velocity dependence, but even Newton's equation of motion, which is commonly regarded as inconsistent with relativity, though it had never suffered any meaningful changes since the 17th century. All revolutions usually try to discard the legacy of the previous generations, and physics is not an exception to the rule. In any case, it was at the beginning of the 20th century that relativity and classical physics parted with each other in the minds of the contemporaries [6] to meet again many years later in the minds of future generations. [7]-[24]. During that period, an invisible and tedious work was under way on digesting special relativity and wondering at its classical roots, which sometimes peeped up like mushrooms from under the ground.

2. What Einstein thought about it

There was at least one scientist in whose mind special relativity was never separated from the basics of classical physics. It was Einstein himself. Newtonian mechanics and Maxwell-Lorentz's electrodynamics had served him well as a basis for the development of his theory of measuring rods and clocks [5]. That's why he never was in position to destroy these branches of science, or give them up. Basing on his postulates, he discovered in them a lot of curious things which until then had remained unknown to everyone including even their great founders.[1] He managed to reconcile the two fundamental notions which seemed to everyone absolutely incompatible with each other. One of them was the all-pervading ether as a playground for electrodynamical phenomena, and the other one – Galileo's principle

---

1. Einstein's intentions with regard to classical physics are felt since the very first sentence of his pioneer work [5]: "It is well-known that Maxwell's electrodynamics – as usually understood **at present** – when applied to moving bodies, leads to asymmetries that do not seem to be inherent in the phenomena." The meaning of the phrase "at present", highlighted here, suggests that those asymmetries could be eliminated by relativity without leaving the realm of Maxwell's theory.

of relativity, which, according to Einstein's endeavor, was extended to electricity and magnetism, as well as to all of nature. Einstein was the first to clarify how these two contradictory notions could get along with each other. He relied on classical physics which had to knock itself out by authorizing unprecedented relativistic effects, such as length contraction and time dilation, which masked the ether from the instruments moving through it by inertia. And so unshakable was his belief in the inevitability and universality of the principle of relativity declared by Galileo that it did not matter to him how odd and inconceivable all these miracles looked and what mechanisms were responsible for their realization in different particular situations. If the principle of relativity required something unbelievable to happen, this something would be sure to happen in full accordance with the laws of nature. But the classical mechanisms underlying these effects sometimes were not evident from the first sight and required tedious work for their exposition. Such was the price paid for the reconciliation between Galileo's relativity and the ability of light to propagate through vacuum in the same way as sound propagates through the air.

Since Einstein declared his postulates for all of nature, he derived the relativistic effects from the postulates and not from mechanics or electrodynamics which, taken by themselves, were not able to provide a universal explanation for the behavior of all the rods and clocks existing in nature. Each rod and each clock could have its own good reasons for behaving in accordance with the postulates. Engaged in general relativity, gravitation and many other cornerstones of physics, Einstein left the investigation of these good reasons to future generations. Only incidentally did he return to special relativity to throw some very interesting and valuable sidelights on his pioneer work [5]. We will dwell on his three most important sidelights which are related to the classical explanations of relativistic effects.

1. How Einstein refined his second postulate.

To discuss this topic, we have first to purify the second postulate from the traditional simplifications practised in textbooks on relativity, the authors of which, in their pursuit of making the material more understandable, often sacrifice the physical meaning endowed

2. WHAT EINSTEIN THOUGHT ABOUT IT 241

to the second postulate by Einstein, and replace this postulate by the law of constancy of the speed of light. Einstein's original formulation reads exactly as follows [5]:

"Every light ray moves in the "rest" coordinate system with a fixed velocity V, independently of whether this ray of light is emitted by a body at rest or in motion. Hence,

$$velocity = \frac{light\ path}{time\ interval},$$

where "time interval" should be understood in the sense of the definition given in section 1".

[According to section 1 of Einstein's work, the two spatially separated clocks, used in this definition, should be synchronized with a light signal sent forth and back between these clocks.]

This is **not** the well-known postulate of the constancy of the speed of light which is equivalent to the following three independent assertions to be observed simultaneously:

1. The speed of light does not depend on the motion of the source.
2. The speed of light does not depend on the motion of the observer, or, much better, on the motion of the instruments used for its measurement.
3. Light propagates forth and back with the same speed.

Only two of them, the first and the last, are contained in Einstein's formulation – the first one explicitly, and the last one implicitly through the reference to section 1. As for item 2, it is not included into the second postulate. Neither can we find it explicitly in section 1 of Einstein's pioneer paper, where the method of clock synchronization is proposed. For that method to work right, only item 3 is needed, and not item 2 which belongs rather to the first postulate as a very important its corollary. When light is used for the clock synchronization, it is only the isotropy of its propagation that is really needed, and not its independence of the motion of the measuring devices.

Why was Einstein so scrupulous in formulating his second postulate instead of assuming just the principle of the constancy of the speed of light in its full and simplest form, widely and sometimes too bravely used in the textbooks on physics: "The speed of light in vacuum has the same value c in all directions and in all inertial reference frames." It is quite brief, sounds so simple, and is certainly valid always and everywhere. What for did Einstein decide to sacrifice its simplicity by eliminating from it the item 2 listed above? Because the message of Einstein's pioneer work was not the principle of relativity – it had been long since proposed by Galileo. Neither was it the ability of light to propagate through vacuum independently of the source as though vacuum was filled with an all-pervading ether – this had been long since proposed by Fresnel and recently explained by Maxwell. The message of Einstein's work was the conciliation between these two well-known ideas which seemed absolutely incompatible with each other, and the postulates were formulated so as to reflect that message. The first idea was expressed in the first postulate, and the second idea – in the second postulate, with reference to section 1 (the relativity of simultaneity) serving as a connecting link between the two – just use the synchronized spatially separated clocks to measure the speed of light involved, and, lo and behold, the ill-fated contradiction between the postulates disappears altogether.

In other words, Einstein's two-postulate presentation would be equivalent to the following simplified triple formulation (which does not refer to the phenomena describable only in terms of quantum mechanics):

1. Newtonian equations of motion are universally valid.
2. The Maxwell-Lorentz electrodynamics is universally valid.
3. The simultaneity of spatially separated events should be established by the light signals.

Here item 1 corresponds with Einstein's first postulate, whereas items 2 and 3 together make Einstein's second postulate. The first two items would be contradictory if not for item 3 which makes the first two get along with each other. On the other hand, the first two items of this formula have survived without any modification, whereas the third item suffered in sequence two refinements discussed below. The first of them was made by Einstein 6 years after

2. WHAT EINSTEIN THOUGHT ABOUT IT   243

his pioneer work and will be given in the next paragraphs while the second one was made by P.W.Bridgman 20 years later than the first and will be discussed in Subsection 3. Both the refinements have remained in shadow and have not become common knowledge.

Unfortunately, the integrity of special relativity and classical physics existed only in Einstein's mind, but not in the minds of his followers. They regarded special relativity as a scientific revolution which swept away classical physics, at least for the speeds of motion approaching the speed of light. But with the old principles given up and the new ones not yet detailed, some physicists and especially philosophers started interpreting relativity in a quasi-scientific way so as to use it as a proof of their current ideology. There appeared so-called conventionalists who tried hard to undermine relativity by declaring all its effects (if not entire science with relativity as an example) to be seeming. They found a loophole in Einstein's method of using light signals for synchronizing spatially separated events and declared all relativistic effects as conventional, i.e. dependent on the definition in a "non-trivial sense". They started from absolutely arbitrary supposition that, in any inertial frame of reference, light may propagate forth and back at different speeds. In due time [25], they introduced a special value ε to represent that difference. According to their scheme, this value resided within the interval $0 < \varepsilon < 1$ whose middle $\varepsilon = 0.5$ corresponded to equal speeds forth and back. That middle was the only possible resort to be left for Einstein's relativity based on the Lorentz transformations. All other values of ε required the Lorentz transformations to be replaced by other ones – much more complicated and anisotropic. All these revised groups of transformations were regarded as simultaneously true because ε proved unmeasurable as well as the one-way speed of light. It was only the round-trip speed of light that made any sense, while the one-way speed of light, as well as all relativistic effects, depended on the convention. If someone, for instance, has measured the length of a moving rod, then, according to the conventionalists, the result should be reported in a very ambiguous way as shown by the following arbitrary example: "The rod in motion is twice shorter than the same rod at rest, given $\varepsilon = 0.5$ or, which is the same, the rod in motion is 3 times shorter than the same rod at rest, given $\varepsilon = 0.789$, or, in other words, the rod in motion is 3 times *longer* than the same rod at rest, given $\varepsilon = 0.0189$, and so on". The list of the options is infinite, which makes the concept of length physically meaningless when applied to a moving rod.

Of course, Einstein did not like it. He immediately responded with a very reasonable and instructive comment [26] where he proposed an unambiguous method of measuring the length of a moving rod *without any light signals* and even *without any clocks*. Let him take the floor.

> "Consider two equally long rods (when compared at rest) $A'B'$ and $A''B''$, which can slide along the X-axis of a non-accelerated coordinate system in the same direction as and parallel to the X-axis. Let $A'B'$ and $A''B''$ glide past each other with an arbitrarily large, constant velocity, with $A'B'$ moving in the positive, and $A''B''$ in the negative direction of the X-axis. Let the endpoints A' and A'' meet at point A^* on the X-axis, while the endpoints B' and B'' meet at point B^*. According to the theory of relativity, the distance A^*B^* will then be smaller than the length of either of the two rods $A'B'$ and $A''B''$, which fact can be established with the aid of one of the rods by laying it along the stretch A^*B^* while it is in the state of rest."

If necessary, this method could be used for synchronizing the spatially separated events or clocks. Let the X-axis be inhabited by an array of independent clocks. Then the two of them that happen to be in the points A^* and B^* could be made synchronous by triggering them at the instants of meeting between the relevant ends of the rods (A' with A'' and B' with B''). Using rods of various length and assigning them different speeds of approaching each other, it is possible to synchronize all the clocks distributed over the X-axis.

However the most convinced conventionalists tried to find a loophole here. To provide identity of the speeds of the two rods moving against each other, those speeds are to be measured, which cannot be done without spatially separated clocks which, in their turn, should be synchronized before Einstein's procedure rather than after it. When reasoning in this way, the conventionalists do not take into account the Einstein's reservation given in the very first sentence of the above citation. It is mentioned there that the equality of the lengths of the rods is to be provided *when the rods are at rest*, which

2. WHAT EINSTEIN THOUGHT ABOUT IT 245

inevitably entails their subsequent acceleration by two identical devices acting in opposite directions. Since the simultaneity of launching the rods is not required, any clocks are excluded from the procedure. Strictly speaking, the rods have also to pass a preliminary test on elasticity, i.e., their ability to follow the Lorentz contraction as exactly as possible. By the right choice of the rods or by improving their material, the accuracy of this metrological procedure can be raised up to the degree determined by the current level of technology.

To summarize this refinement, let us include it verbally into the second postulate. Without the refinement, the second postulate can be paraphrased as follows:

> Light propagates forth and back with the same speed. This speed is independent of the motion of the source, provided it is measured by two spatially separated clocks which are synchronized with a light signal sent from the first clock to the second one and then returned back.

The physical meaning of this formula is exactly the same as that given by Einstein and cited above on page 241. But the wording is different, which spotlights the origin of the loophole used by conventionalists – the speed of light plays the two interrelated roles there. It is simultaneously an object of measurement and a means for adjusting the measuring instruments. With taking Einstein's refinement into account, this interrelation is eliminated:

> *Light propagates forth and back with the same speed. This speed is independent of the motion of the source, provided it is measured by two spatially separated clocks which are synchronized with two identical rods, shot against each other by means of two identical catapults.*

The next and final edition of this postulate, given in Subsection 3, will be much shorter.

2. What Einstein thought about the ether.

Before developing the theory of general relativity, Einstein was inclined to give up the ether as a useless notion. In his pioneer work [5] he left no doubt about it:

> "These two postulates suffice for the attainment of a simple and consistent electrodynamics of moving bodies based on Maxwell's theory for bodies at rest. The introduction of a 'light ether' will prove to be superfluous inasmuch as the view to be developed will not require a 'space at absolute rest' endowed with special properties, nor assign a velocity vector to a point of empty space where electromagnetic processes are taking place".

But the deeper was Einstein penetrating the bowels of general relativity and gravitation, the stronger was his conviction that the emptiness as a background for all physical objects is a hardly digestible thing from both physical and philosophical points of view. Sixteen years later, this brought him to the following conclusion [27]:

> "More careful reflection teaches us, however, that the special theory of relativity does not compel us to deny ether. We may assume the existence of an ether, only we must give up ascribing a definite state of motion to it, i.e. we must by abstraction take from it the last mechanical characteristic which Lorentz had still left it. We shall see later that this point of view, the conceivability of which I shall at once endeavor to make more intelligible by a somewhat halting comparison, is justified by the results of the general theory of relativity."

And further:

> "To deny the ether is ultimately to assume that empty space has no physical qualities whatever. The fundamental facts of mechanics do not harmonize with this view. For the mechanical behavior of a corporeal system hovering freely in empty space depends not only on relative velocities, but also on its state of rotation, which physically may be taken as a characteristic not pertaining to the system in itself. In order to be able to look upon the rotation of the system, at least formally, as something real, Newton objectivizes space. Since he classes his absolute space together with real

2. WHAT EINSTEIN THOUGHT ABOUT IT

things, for him rotation relative to an absolute space is also something real. Newton might no less well have called his absolute space 'Ether'; what is essential is merely that besides observable objects, another thing, which is not perceptible, must be looked upon as real, to enable acceleration or rotation to be looked upon as something real."

In other words, Einstein regards the ether as a privileged entity from which accelerations, but not velocities, should be counted and which remains undetectable as long as everything moves just by inertia (including the free fall within a small region of space).[1] It does not matter what we call it. Newton had called it "absolute space". Fresnel, Maxwell and Lorentz called it the ether, though they erroneously regarded it as an object from which constant velocities and free fall accelerations could be counted. (It is this property that should be taken away from their ether to make it admissible as a physical reality.) Some physicists call it "vacuum" (or "physical vacuum" so as to impart a splash of materiality to the emptiness). John Archibald Wheeler and many other physicists call it spacetime to stress the mutual dependence of space and time, discovered by Einstein. As for the invisibility of this ether within the limits of special relativity, that invisibility is not so unusual as it might seem at first sight. Such a "well-known" and absolutely corporeal notion as the inertial mass also remains insensible as long as we consider the motion by inertia. But this is not a good reason for calling it superfluous and expelling it from the vocabulary of physics.

The choice of a proper term for emptiness might depend on the topic involved. In the non-postulated relativity the term "ether" would be perhaps preferable because it suggests continuity between the pre-Einsteinian and post-Einsteinian physics.

---

1. Since in the general theory of relativity the laws of nature are usually presented in an invariant four-dimensional form, this sometimes leads to a confusion. The invariance of the mathematical equations is erroneously interpreted as the physical invariance of the laws of nature, which would be incompatible with any ether. Einstein clarified this situation in [28]: "The fact that the general theory of relativity has no preferred spacetime coordinates which stand in determinate relation to the metric **is more a characteristic of mathematical form of the theory than of its physical content.**" One may find more details about this in [29].

3. Uniting the past and the future.

When in 1951 Einstein was giving the author's post-estimation of special relativity in his *"Autobiographical Notes"* [30], he made a very remarkable comment:

> "The universal principle of the special theory of relativity is contained in the postulate: The laws of physics are invariant with respect to Lorentz transformations (for the transition from one inertial system to any other arbitrarily chosen inertial system). **This is a restricting principle for natural laws, comparable to the restricting principle of the nonexistence of the *perpetuum mobile* that underlies thermodynamics**." (Highlighted by L.L.)

When solving a routine problem in mechanics, we normally have a choice whether to use the equations of motion in order to arrive at the desired solution, or to prefer the laws of conservation, which would bring us to the same result. But in special relativity we are accustomed to start all considerations from Einstein's postulates, i.e., from a restricting principle which plays the same role as a conservation law. Moreover, we are usually doing it in a blindfolded manner being unaware of other ways which also exist, or, all the worse, we sometimes deny the very existence of other ways. Of course, on the frontiers of science, such a choice is hardly realizable because the knowledge and understanding are too scarce there − the beggars can't be choosers. But in such mature fields as classical electrodynamics and mechanics this choice is always possible, though sometimes we have to apply ourselves to find the ways. On the next pages of the same book [30], Einstein explains his attitude toward these ways. He finds the search for them not only desirable but even obligatory:

> "First a critical remark concerning the theory [of special relativity] as it is characterized above. It is striking that the theory (except for the four-dimensional space) introduces two kinds of physical things, i.e., (1) measuring rods and clocks, (2) all other things such as the electromagnetic field, the material point, etc. This, in a certain sense, is inconsistent: strictly speaking, the properties of measuring rods and clocks (as objects consisting of moving atomic configurations) should be derived from the basic equations, instead of being handled as theoretically self-sufficient entities. However, we still have to concede this inconsistency because, as was clear from the very beginning, the postulates of

2. WHAT EINSTEIN THOUGHT ABOUT IT

the theory are not strong enough to deduce from them equations for physical events sufficiently free from arbitrariness in order to base upon them a theory of measuring rods and clocks. If one did not wish to forgo a physical interpretation of coordinates completely (something that, in itself, would be possible) it was better to permit such inconsistency − **with the obligation, however, of eliminating it at a later stage of the theory**. But one must not legitimize the sin just described so as to imagine that distances are physical entities of a special type, intrinsically different from other physical variables ('reducing physics to geometry', etc.)"

The phrase about the obligation was highlighted by L.L. The story about fulfilling that obligation will be given in the next section. As for "legitimizing the sin", this, unfortunately, happens rather often in the textbooks when the authors prefer to explain relativistic effects only by the new properties of space and time discovered by Einstein. One has a chance to be taught, for example, that "the moving clock must tick more slowly **because time is Einsteinian and not Newtonian**", which sounds rather authoritative. This wording, supported by mentioning a lot of well-known experiments, proves so persuasive that the search for other ways of explanation seems either hopeless or at least excessive. And it does not occur to a reader that the two times − Newtonian and Einsteinian, are identical, provided the properties of space and time are identified with the behavior of rods and clocks (at least for the time being until the ether, or space-time, or whatever we call it becomes available for ascribing these properties directly to it), and the laws of classical physics are used exactly in the form offered by Newton, Maxwell and Lorentz without any simplifications or corrections. The reader can imagine a clock of any construction (for example, two parallel mirrors with a light flash reverberating between them), set it mentally in motion with a constant velocity, move mentally himself or herself to the end of the 19th century, and calculate the tick of this clock without any relativity. He or she can make sure that the result of such mental excursion will be in accord with relativity and that the speed of motion, however close to the speed of light it might be, does not put any limit to the range of validity of classical physics. As for the ether, it will fade away as soon as the reader imagines a pair P of spatially separated clocks, moving through the ether (and "damaged" by the ether), and estimates the tick of a clock fixed to the ether "in the eyes" of P. As we know from the preceding subsection, the ether, even having faded away, proves not so excessive as we are often taught. Even when we face the situation in which all the bodies are moving by inertia, the "undetectable" ether still makes us conscious of being in the privileged position from which accelerations should be counted.

To appreciate the practical importance of the obligation which, according to Einstein, all of us are under, let us consider, for example, a straight, long current-carrying wire. When such a wire is at rest, it is electrically neutral. There is only a magnetic field and no electric field around it. But if this wire is set in motion in the longitudinal direction with a constant speed, then, according to special relativity, there appears an additional electric charge within the wire, which creates the electric field in the surrounding space directed radially − perpendicular to the motion. The signs of the charge and of the electric field depend on the direction of the current with respect to the direction of motion − whether these two directions are the same or opposite. If a question arises about the origination of this additional charge − where does it come from − usually the answer follows that it comes from infinity, as the ideal linearity of the current can be achieved only in an infinitely long wire. No one doubts that the additional charge will appear indeed because its mysterious appearance is sanctified by the high authority of special relativity. This answer is correct, though it remains unclear how this infinity is arranged to generate this charge. Let us now deprive this wire of infinity by rolling it up into a current-carrying ring rotating round the axis of symmetry of the circular current. Where does the charge come from now? The right answer is that now there arises no additional charge at all, and no electric field, given the wire is thin enough.

Special relativity is not valid here because the motion is rotational, which takes this case out of the competence of this theory.[1] But there are also other cases of rotational motion (for example, the

---

1. Sometimes, though, one can come across other opinions about this situation. Some authors believe that the excessive charge still exists, though they call it "fictitious", "effective", or "apparent", which, somehow, does not prevent these authors from regarding the electric field, created by this charge, as pretty real. So high does relativity stand in the eyes of physicists and even some engineers that they are ready to believe in the existence of the electric field whose lines of force originate from and terminate in a fictitious charge, that is, actually nowhere.

2. WHAT EINSTEIN THOUGHT ABOUT IT

same rotating ring as above, but with a radially polarized dielectric, which acts as though filled with stationary magnetic dipoles creating a magnetic field in the surrounding space) to which special relativity can be applied with fair accuracy. We wonder how to know when special relativity is applicable, at least approximately, and when it responds with a wrong answer. As both the examples have been taken from electrical engineering, we have to concede that, in spite of the impressive age reached by special relativity and the well-known revolutionary changes caused by this theory, we are still unaware of how far the limits of its validity extend. We feel embarrassed and often respond with a wrong answer when asked about such a "detail" as reversibility of relativistic effects. If, for example a rod has been set in motion with a uniform velocity and has undergone a Lorentz contraction, what can we say about its length after it is stopped and returned to the state of rest again? Does it have now exactly the same length as it had at the beginning of this procedure? The right answer is "No, it does not". It has suffered a residual deformation as though it were deformed by heating and then returned to the original temperature. To restore its former length exactly, the rod must be absolutely elastic. Some people at first disagree with such an answer. They know that the deformation of the rod both forth and back takes place in accordance with the Lorentz transformations, which are absolutely symmetrical and seem to exclude any residual deformation. Rarely does anyone try to answer this question without relativity, or even think seriously of such an option. The more we think about such examples, the more clearly we see that in special relativity and about it, even now, on the eve of its hundredth anniversary, there are many important options not yet mastered by physics. As for the epoch of the 50th, which Einstein's declaration belongs to, it was the time of the first conscious attempts to pave a classical way to special relativity. As you will see from the next section, these attempts were stimulated not so much by Einstein's appeal as by the desperation for getting an answer to the childish questions about the physical origination of time dilation and other relativistic effects, which seemed so mysterious and non-tangible.

3. Groping ways to non-postulated relativity

As was mentioned at the very beginning of this historical review, the initial part of the obligations, declared by Einstein in his *Autobiographical Notes* had been fortunately fulfilled by Lorentz and FitzGerald as early as before 1905. It was then that the contraction of moving bodies and the mass-velocity dependence were discovered. But Einstein's main innovation − the behavior of the moving clocks, and especially the spatially separated clocks responsible for the relativity of simultaneity − stood opposed to classical physics even in the middle of the 20th century. Therefore the first achievements of classical physics in explaining relativistic effects (such as the length contraction) were qualified as a fallacy connected with the ether. Einstein's refinement of his second postulate, considered in the previous section, was only the first step in erecting the bridge between classical physics and Einstein's relativity. But unfortunately, this step remained almost unnoticed not only by the opponents of relativity, who criticized it as having a "loophole", but also by the proponents of Einstein's theory, who did not take this refinement seriously in view of the evident triumph of relativity in physics. Therefore, the further work on digesting special relativity was the destiny of single, independent and insistent personalities with a strong desire to dig out the lacking information and indifferent to how other people would estimate their activities. P.W.Bridgman was one of them [7]. He felt some incompleteness in using light signals for the clock synchronization and was looking for other ways. At last he arrived at the method which is so simple and natural that hardly would be ever overridden by anything else. It turned out that in order to synchronize the spatially separated clocks we have to do ... nothing. If two clocks are synchronized at the same point of space and one of them is just transported slowly and carefully to another destination with a speed far enough from that of light, the hands of this clock will adjust themselves automatically, and, at the new location, the clock will show the same time as if it were synchronized by light signals according to Einstein's original proposal. On a TV screen located midway between the clocks, the readings of the two clocks will be identical. They would remain identical even if the above procedure was made on a speeding platform. But, according to the instruments which were at rest, these clocks would become non-synchronous. The clock that was transported is responsible for it. The

3. GROPING WAYS TO NON-POSTULATED RELATIVITY

good reason for this non-synchronism is the slowdown (or speeding up) of the tick of the clock in the process of its transportation. Bridgman knew nothing about this reason, which made his consideration rather sophisticated [7]. He based his consideration on the principle of relativity. If the transported clock did not behave so "strangely", then the observer sitting on the platform and looking at the TV screen midway between the clocks would see the readings of the clocks become different there, and that difference could be used to detect the absolute motion of the platform.

Bridgman's proposal was actually the next methodological refinement of the second postulate. Though he never looked at his achievement from this standpoint, his proposal is suggestive of being included into the second postulate instead of a reference to either the light signals, used in Einstein's original formulation [5], or moving rods, proposed by Einstein later [26]. This would make Einstein's argumentation against the conventionalists even stronger and the postulate itself much shorter:

Light propagates forth and back with the same speed which is independent of the motion of the source.

The procedure of clock synchronization had been removed from the postulate as superfluous − the clocks did not need any readjustment but a mere transportation to their destinations without occasioning them any damage. Neither is it mentioned there that the speed of light is independent of the motion of the observer. Reminding it would not be against the truth − it would be against Einstein's endeavor.

The next enthusiast who made a significant contribution to the understanding relativity was H.E.Ives − a famous experimentalist. Together with G.L.Stilwell, he is widely known as the author of the first man-made experiment which demonstrated the slowdown of the tick of a moving clock and, in contrast to Michelson's experiment (and most other experiments supporting special relativity at the time), was expected to give, and actually gave, a positive outcome [8]. But very seldom is it mentioned that Ives himself regarded his experiment as a proof of the existence of the ether and hence, as he erroneously supposed, a disproof of relativity. He was discouraged by the fact that the scientific community had interpreted his experiment in the way exactly opposite to his expectations. So he turned to the theory and published a set of thoroughly prepared articles almost unknown nowadays [9]–[12], where he tried to prove the existence of the ether and to "disprove", as he thought, relativity. "What else but the ether can make the moving clock slow? If the clock is moving through emptiness, i.e. through nothing, and has no interaction with any other body, then what non-earthly force makes it slow?" – this is a brief paraphrase of Ives's quite reasonable considerations regarded everywhere as strange, archaic, and not worthy of mentioning among the savants. It proved not so easy to figure out that both approaches – the one based on the ether and the other one based on relativity – gave the same results and did not contradict each other, because the existence or non-existence of Lorentz's ether does not affect the eventual result of any investigation.

The ultimate result of Ives's theoretical investigation was the clarification of the fact that not only the length contraction and mass-velocity dependence, but also time dilation are derivable from classical physics. But this was not recognized widely by the scientific community, because classical physics and relativity were kept in Ives's mind opposed to each other, and he could not imagine them predicting the same result. Since he was preoccupied with classical reasoning being in accord with the result of his experiment, he erroneously regarded special relativity disproved. As for most of the members of the scientific community, their minds were preoccupied with relativity (being also in accord with the result of Ive's experiment), which made them regard Ive's experiment as one more evidence against classical physics. That's why Ives's reasoning was not accepted even in those parts which could be qualified now as an unnoticed milestone on the way to non-postulated relativity.

4. Building bridge to classical physics

One more contributor who pioneered non-postulated relativity was L.Janossy, a European physicist. He summarized the results of his groping in a vast article [14] (1957, in Russian) and a book [15] (1971, in English), leaving alone his early article [31] (1952 in Hun-

4. BUILDING BRIDGE TO CLASSICAL PHYSICS 255

garian). He approached special relativity from another side. To clarify the good reasons responsible for a certain relativistic effect taking place in a certain system, for example in a solid body, he considered the origination and evolution of that effect after the system was set in motion with a constant velocity under the action of any force but gravitational. It does not matter how the velocity of the body is changed, because the final state of the body is determined by the final value of its velocity. The body may be thought of even as undergone an instantaneous acceleration without any deformation. However this non-deformed state is unstable. Now the particles of the body are to find their new equilibrium positions predetermined by the Lorentz transformations.

When such jump in velocity takes place, the frame of reference which accompanies the rod, or the clock or another similar system involved, also undergoes this jump and hence is not inertial. Therefore not every system involved is obliged to obey the Lorentz transformations and to behave as described above. Only ideally elastic systems (or "connected systems" as Janossy calls them) do behave in accordance with the Lorentz transformations. As for the ordinary elastic systems, they do it only approximately – with a certain error derivable only from the basic equations, e.g., from classical mechanics. But very rarely does Janossy track down the entire cause-and-effect chain between the classical equations and the relativistic result. Instead, he often uses relativistic explanations, though based on his own interpretation of the Lorentz transformations – different from that given explicitly by Lorentz or Einstein. To catch the point of it, let us take for example the process of electrical polarization of a moving magnetic dipole (that was considered in Section 2.8.)

Imagine a rectangular conductive current-carrying frame with its axis of symmetry which lies in the plane of the frame and is parallel to any two opposite conductors of the frame. When such frame is set in motion along the axis of symmetry, the conduction electrons within the frame are redistributed between the two longitudinal conductors, so that the frame acquires electrical polarization directed perpendicular to the motion. As shown in Section 2.8, this redistribution, usually derived from special relativity, can be explained in terms of electromagnetic induction. As for Janossy's publications, it is hidden there deeply inside the formalism based on the rule of relativistic addition of velocities. This rule is deduced, though, not from Einstein's postulates. The role of the postulates is played by the

Lorentz transformations interpreted as equations which govern the relativistic deformation.

Janossy's approach to the Lorentz transformations is nothing else but the third possible way of interpreting the physical meaning of the primed and non-primed variables in these transformations. The first way belongs to Lorentz who regarded the non-primed variables as true coordinates of a certain physical system in space and time, while the primed variables were fictitious and served only as means for solving the Maxwell equations. Einstein interpreted the non-primed variables in the same way as Lorentz, but the primed variables were declared by him as absolutely real – they indicated how the non-primed variables looked like in the eyes of the instruments which accompanied the object of their measurement. As for Janossy, he interpreted both kinds of the variables as obtained by one and the same set of the measuring instruments which is always at rest and is applied to one and the same body – first before the acceleration of this body (primed variables), and then after the body is accelerated and all its particles have found their new equilibrium positions (non-primed variables).

Einstein knew about such a possibility and sometimes used it (suggesting for example the clock paradox). But he did not formulate it explicitly in the form of a general rule because it would mean for him a considerable loss in generality. Einstein was a discoverer and the global generality was much more important to him than particular comprehensibility, whose further development was left by him to the future generations. As for Janossy, it was comprehensibility that he was striving for, and his interpretation of the Lorentz transformations was a good methodological and pedagogical finding. But from the standpoint of his endeavor, it would be better for him to go even further and start his explanations from classical physics in the way we did in Section 2.8. If someone asks you about the reason for the electric polarization of a moving magnetic dipole, it will be much more understandable to him or her to base your explanation on the law of electromagnetic induction rather than on the third possible way of interpreting the Lorentz transformations. Though Janossy had done a good job to revive relativistic effects, the bridge connecting special relativity with classical physics still remained unfinished. If Janossy had paid a due attention to Ives's classical derivations, which inadvertently unified relativity with classical physics, he would somehow have adapted them to his dynamic approach, so that the last span of the bridge would have been erected. But how could Janossy figure out that the information he needed was hidden in the considerations whose announced purpose was to disprove relativity?

Janossy's ideas were favorably accepted in the community of physicists of the former USSR. They were published there [14] and detonated a lively controversial discussion, which lasted for years and eventually was put to order by a very instructive and detailed article published by E.L.Feinberg [20]. In contrast to his predecessors in this field, E.L.Feinberg was not striving for figuring out the physical mechanisms of all the main relativistic effects because he had no doubt in their existence. Very thoroughly and patiently, with a lot of instructive remarks and reservations, he explained that if mankind were so stupid that relativity had not been created either in 1905 or in the following decades [1], the scientists would all the same draw the relativistic effects from the bowels of classical physics just in the same way as Lorentz had discovered the length contraction and mass-velocity dependence. Feinberg admitted that there are situations when the dynamic approach is not only possible but even obligatory. But on the other hand, Einstein's postulates are preferable when everything is stationary and it is the properties of spacetime that come to be under investigation.

Feinberg's article was an eye-opener for the people striving to understand relativity instead of memorizing its rules and turning them into a custom. It also established guidelines for publishing new textbooks on special relativity where the dynamic approach would play a leading role in explaining the main relativistic effects. As for the particular techniques for doing so, many of them could be borrowed from the works of Lorentz, [2], Bridgman [7], Ives [9]–[12], Janossy [15], Feinberg [20], and other contributors [3],[4],[13], [16]–[19], etc.

But there remained still some ways to be groped between classical physics and relativity. The mechanism of the electric polarization of a moving magnetic dipole with its magnetic moment perpendicular to the motion (given above as an example) had not been clarified as yet. This electric polarization had been deduced from Einstein's postulates (in most textbooks), from the relativity of simultaneity [32], or

---

1. In other words, if Einstein were born a few decades later.

from the Lorentz transformations interpreted in a way different from that of classical physics [15], but never directly from the Maxweell-Lorentz electrodynamics or Newtonian mechanics. Even the most experienced authors [13] failed to find the way of doing it and conceded classical electrodynamics to be helpless there without relativity. Besides, there remained a much more important gap that deserves a special consideration.

5. Closing the gaps

A) *Slow transportation of a clock along a speeding platform*

When thinking of the Lorentz length contraction, we usually associate it with a rod or a ruler which becomes shorter as soon as it is set in motion. This simple example is exhaustive in visualizing this effect. As for the time dilation, it does not prove so simple. At first thought we may associate it with a clock whose ticking becomes slower as soon as it is set in motion. The good reasons for such behavior depend on the clock's design and can be explained in terms of classical physics unless the action of the clock is based on quantum mechanics. But at the second thought we realize that we miss the time-space dependence – this main innovation of special relativity. This dependence could be envisioned by a slow transportation of the clock along the speeding platform. The rate of its ticking (its frequency) after the transportation will remain the same, but its reading (its phase) will be changed in accordance with the place where the clock is brought to. There is a good reason for this change: during the transportation, the clock is moving a little bit faster or more slowly than the platform, which makes the rate of its ticking a little bit slower or faster respectively. That little bit, which could be made as small as one likes, multiplied by the time of the transportation, which proves correspondingly as great as necessary, will result in the ultimate change in the clock's reading, which does not depend on the speed of transportation unless it is commensurable with the speed of light. This explains the relativity of simultaneity in terms of classical physics, given the reasons for the slowdown of the tick of the clock in those terms are already known. This consideration is OK until the motion of the platform is stationary – without any change in its

5. CLOSING THE GAPS

velocity. If we want to apply it to the interpretation of relativity as based on the dynamic approach, this explanation does not work, and we are to search for something else as follows below.

B) *Dynamics of spatially separated clocks*

When the platform is at rest, time doesn't depend on space and is the same everywhere. The dependence of time on the longitudinal coordinate x is described by a straight line parallel to the x-axis. But when the platform is set in motion, this straight line inclines to the x-axis by a certain angle determined by the speed of the platform. This time-space dependence might be associated with an array of clocks fixed to the platform and aligned in a single file along the motion. Since the time-space dependence is linear, it is sufficient to consider just a pair of such spatially separated clocks. Let them show the same time before the start of the platform. What will they show after the platform has been set in motion and is moving with a constant speed? Whatever might they show, their indications will be identical because they are accelerated in the same conditions. But being in motion and showing the same time (according to the clocks at rest), they are no more synchronous. If slowly brought together, they will show different time. These clocks are wrong and cannot serve as measuring instruments. The same happens to the rod broken into two fragments. (Janossy calls such systems disconnected). Let us join these fragments loosely one after the other. When set in motion, each of these parts of the rod experiences duly the Lorentz contraction. But the distance between the middle points of the fragments remains the same. This leads to a gap between the fragments, which are no more in touch with each other. To make the rod contractible, we have to glue the parts firmly together.

The same thing happens to the spatially separated clocks. To make them obey the Lorentz transformations, we have to combine them into one equilibrium system. We can do it in many ways with the same result. The simplest of them is to connect the two clocks with a continuous electromagnetic wave. Let one of them be a master clock having its own mechanism to drive its hands and to send an electromagnetic wave to the second clock which doesn't generate the signal of its own and just repeats or amplifies the received signal turning it into the movement of the hands. What will happen to these clocks when the platform they are attached to is set in motion? The master clock will slow down its ticking, and so will the slave clock but not immediately. This clock obeys the electromagnetic wave which is continuously coming to it. For a certain time, necessary to the wave for covering the distance between the clocks, this wave is "unaware" of the motion of the platform. While propagating through space, the pulsations of the wave do not know anything about the clocks and their motion. As for the slave clock, now it is moving either in the same direction as the wave or against it, so that the pulsations of the wave reach this clock more rarely (or more often) than before the acceleration. This makes its hands move respectively in a slower or faster tempo during a limited interval of time whose duration is equal to the time it takes the wave to cover the distance between the clocks. During this interval, the slave clock is receiving the pulsations of the wave which at the moment of acceleration were on their way between the clocks. This will ultimately make the slave clock either slow or fast with respect to the master clock. If in addition to this first-order effect, we also take into account the contraction of the distance between the clocks, we will see that the difference in the readings between the two clocks will be exactly as predicted by the Lorentz transformations. Thus, it is the electromagnetic pulsations, being on their way between the clocks at the moment of the acceleration, that play the main role in visualizing the relativity of simultaneity as a physical reality. They do their job in full accordance with Einstein's postulates.

This might be summarized in the following way. A set of test objects, sufficient for envisioning the main relativistic effects in their dynamics, must include not only rods and not only clocks isolated of each other but also a pair of continuously synchronized, spatially separated clocks, attached to a common plate. When set in motion, the rods become shorter, which illustrates the length contraction, the clocks slow down their ticking, which demonstrates the slowdown of all the processes that take place on a speeding platform, and the pair of connected clocks changes their relative readings, which helps us to envision the time-space dependence – the main innovation of special relativity. Normally, a similar set of objects is necessary to provide a set of measuring instruments.

It should be noted that every real rod is populated with a lot of spatially separated clocks continuously synchronized with each other.

5. CLOSING THE GAPS 261

Every atom within the rod is a clock. If all these clocks were not interconnected, the timing between the parts of the rod would be violated, so that the Lorentz transformations would be invalidated. Even the living beings consist of a lot of clocks linked closely to each other. The dynamic approach to the relativity of simultaneity is especially remarkable when applied to such beings (or their mental models) – especially to those of them who are very extensive, such as crocodiles or snakes. To produce a noticeable effect, they must be monsters of fabulous length. Let us imagine a crocodile which is so extensive that it takes a year for light to cover the distance between its head and the end of its tail. What will happen if such a crocodile is set in motion (by any force but gravitational) with a speed close to that of light?[1]

Let us disregard the enormous forces of inertia within crocodile's body, which may be caused by the acceleration and have nothing to do with special relativity. It is not they that we are interested in. We may even aggravate the abstractness of the picture by assuming an instantaneous setting of the crocodile in motion with all the particles of its body remaining at first in the same places as before the acceleration.[2] After instant acceleration, it takes a considerable time for all the particles in the crocodile's body to find and occupy their new positions of equilibrium and to reset the phases of their electrons rotating around their nuclei. The duration of the transient cannot be shorter than it takes light to cover the entire length of the body from the head to the end of the tail. That duration may be even many times longer, so that all the particles would be able to communicate with each other either directly or through mediators.

As soon as this active exchange of information is over, the tail of the crocodile will become either older or younger than the head in accordance with the direction of the crocodile's motion – whether the

---

1. The answer to this question was giv4en in Section 2.7. (See Fig.21...Fig.24 on pages 184-187 with the relevant explanation.) We reproduce this answer here once again so as to update that story without distracting the reader from our review.
2. If someone regards such aggravation indigestible, it might be a good consolation to assume that the crocodile is just a product of cybernetic imagination. We can even imagine that, instead of setting the crocodile in motion, a new crocodile is instantaneously constructed on the speeding platform as an exact copy (atom for atom) of our crocodile for a certain moment of time (as estimated by the instruments which are at rest).

head moves ahead of the tail or the other way round. All the other internal and external parts of the crocodile's body will be of intermediate age between the head and the tail according to their positions. Of course, it is supposed that the master clock is situated somewhere within the crocodile's brain, whereas all the other clocks within the body are controlled by the master clock. It is important that not only the instruments at rest but also the instruments which accompany the crocodile would be able to register this paradoxical ageing unless they "oversleep" the transient or forget their former readings.

For the sake of certainty, let us assume that the setting in motion takes place in a tail-forward fashion so that the tail proves younger than the head. If the crocodile decided to smell and investigate the tip of its tail, a very spectacular scenery would expand before the instruments at rest. While the crocodile is slowly bending its tail, the tip of the tail slowly traces out an arc which starts at its original position and ends up in the monster's nose. During that travel, the tip of the tail is ageing right before the eyes and ends in exactly the same age as the crocodile's nose. So, the result of smelling is predetermined – the crocodile will never learn that its tail proved younger than its nose, unless the monster had been advised to investigate the records of the observations taped by the instruments attached to its tail during the transient and to compare them with the similar records registered by the instruments attached to the crocodile's nose. Those records would cause a lot of surprise in the crocodile's mind.

To make the end of the story as impressive as in the case of the twin paradox, let the crocodile happen to be a female which laid an egg at the last minute before the acceleration and fixed that egg to the platform at the tip of its tail. Unlike the tail, this egg is not connected to the crocodile's body in any way, and hence its ageing during the transient will be of the same rate as the crocodile's brain. If shortly after the acceleration a baby crocodile hatches out of that egg, the rate of its ageing will be also the same as that of its mother's brain because the baby is an independent being and has never left the place of its birth. This means that, during the transient, the baby will be ageing much faster than its mother's tail, whose ageing will be retarded for the reasons discussed above. If the speed of the platform approaches that of light, the life within the tail during the transient will almost die away and the duration of the transient will increase correspondingly. This may bring us to a situation in which

5. CLOSING THE GAPS

the "baby", at the end of the transient, may prove even older than its mother's tail, but still younger than the mother's head. When we think of the baby's age, we imply only the age of its head (where its master clock is supposed to be located), and not the distributed age of its growing body, which will be left out of the scope of this story. The master clock of the baby is supposed to be triggered at the moment of the baby's emergence from the egg.

Now, there comes a payoff of this scenario. The mother brings again the tip of its tail to its nose, and the baby takes a chance to have a ride on it – with the baby's head pinned to the tip of mother's tail, and the baby's body being dragged behind. During this "ride" the baby, in spite of its independence, is ageing with the same rate as the tip of the mother's tail. During the ride, both the baby's head and the mother's tail have higher rates of ageing with respect to the mother's head, though the reasons for these increments in the rate are different – the tail is ageing faster because the controlling signals from the mother's brain come sooner during the ride, while the increment in the rate of the baby's ageing is due to the baby's moving a bit more slowly than the platform. (If these two increments had been not the same, than an observer fixed to the platform would have been able to detect the absolute motion of the platform.) Since before the ride the baby is older than its mother's tail, after the ride it will become older than its mother's brain – the same result as in the twin paradox where a point-size mother makes a round-trip space travel to become younger than her son.

C) *The limits of special relativity*

It is high time now to make an important reservation related to the dynamic approach and its range of validity. For the extended bodies to behave like the mother crocodile in the above scenario, they must be ideally elastic in an unusually broad sense of the word. All their particles must not only have positions of a stable equilibrium, but also must be continuously synchronized with each other in all their rotations and other regular movements. Otherwise, they may disobey the Lorentz transformations, and their particles may behave like the baby crocodile in the above scenario. Imagine for example that, before the acceleration, the mother crocodile was hungry and ate up an indigestible alarm clock. After the acceleration and transient are over, that clock will show not the local time of the stomach where it will happen to be at the moment, but rather the local time of the crocodile's brain or, more exactly, the time shown by the crocodiles master clock, wherever it might be located.[1]

This is valid not only for living beings or their imaginary models, but also for all inanimate objects. If we take for example a rod and set it in motion, its particles must not only find their new positions of equilibrium, but also establish new equilibrium phases of their internal periodical motion. Stable time zones, continuously changing along the motion, must be formed inside the rod. These time zones will be detectable not only by the clocks at rest but even by the clocks in motion, unless the latter "oversleep" both the period of acceleration and the subsequent transient. This is a special kind of elasticity which is not typical in common practice. It might be called a time-space elasticity.

But do such ideally elastic bodies exist in nature? Of course, not. All that we have in nature is at best only an approximation to them. The absolute elasticity, leaving alone the time-space elasticity, exists only in our mind, that is, in abstract images we commonly use in science. All the real bodies, after removing the cause of their deformation, restore their original condition only approximately, though in the physical theory, they may be thought of to be as close to the ideally elastic as one likes. All the more this reservation proves effective in living beings, that are built of not only elastic but also amorphous materials, leaving alone the biological cell, which sometimes may be inclined to acquire an autonomy from the master clock located somewhere too far away.

Does this mean that special relativity is valid only approximately? It depends on how we use it. The postulates are always valid. The Lorentz transformations are also exact as long as we use them for solving the Maxwell equations or to see how a phenomenon

---

1. If we do not know the whereabouts of the master clock within crocodile's body, we can locate it by dispensing a lot of independent clocks to different parts of the body before the acceleration, and comparing their readings – after the transient – with the readings of the biological clocks belonging to the nearby tissues. The coincidence between the readings of the foreign and the aboriginal clocks will expose the desired tissue which contains the master clock.

5. CLOSING THE GAPS 265

looks like if observed from different inertial frames of reference. They remain exact even if the object of measurement is slowly changing its position in the moving inertial frame of reference – a rod or a magnetic dipole is slowly turning, or a clock is being slowly transported along a speeding platform. But the Lorentz transformations do become approximate and may even fail when a change in the uniform velocity of motion happens to the object of observation or to the measuring instruments. Let us take, for example, a rod which is made of an amorphous material and is therefore absolutely plastic. If even, for the sake of clarity, we assume that all the particles of the rod are set in motion instantly without changing their positions (so that all inertial effects prove insignificant), these old positions will prove equilibrium – manifesting the absolute plasticity of the rod. Such rod would prove essentially longer than predicted by the Lorentz transformations. If even the rod is made of elastic material, in real life it bears some traces of plasticity and, being first heated and then cooled up to the initial temperature, does not restore exactly its former shape and size. Having been set in motion and then stopped, it will suffer a small residual deformation. The same refers to a clock, though it looks less vulnerable to this effect because, unlike the rod, the clock can be made as small as necessary. This remark can be extended even to the system consisting of two point-size spatially separated clocks, continuously synchronized by an electromagnetic wave. Such system could be thought of as absolutely elastic if it were not for the plate which the clocks are fastened to.

How could we estimate the error caused by the plasticity of real objects or of the measuring instruments? The relativity based on Einstein's postulates, is helpless against this problem. This job is just within the competence of non-postulated relativity, whose support proves not only possible but even obligatory here in full accordance with Einstein's advice cited above.[1]

The kind of elasticity we are concerned with is a much more general notion than the ordinary elasticity as used in mechanics. It refers even to such a non-classical object as a single electron. Let us place an electron in the electric field which is so strong that the electron is accelerated up to the speed of light earlier than its mass has time to

---

1. Such support has been implicitly considered in the literature under the name "Lorentz boosts" and proved effective indeed.

grow and prevent the electron from overcoming the light barrier. Will this electron reach the speed of light in vacuum? "Surprisingly", it will. How strong must be that field? An approximate calculation shows that it must be close to the electric field on the "surface" of the positron which "comes in touch" with the electron, both of them regarded as classical spherical objects of the size about their classical radius. Their contact with each other brings about the well-known result. Both of them successfully reach the light barrier and convert into a pair of photons, which gladly fly away with the speed of light.

What has been said about a suddenly accelerated object refers also to the measuring instruments by means of which this object is observed. In the common explanations of the relativistic effects, these instruments sometimes remain in shadow – hidden under the alias "observer", which, at the first thought, is normally associated with a human (or at best a robot) who is standing like a sentry on duty looking through a binoculars at a moving frame of reference. This image proved so traditional that it was a kind of revolution when J.Terrel [33] unexpectedly discovered that what the observer saw through his binoculars might not undergo the Lorentz contraction, and some scientists even thought that that contraction proved not only invisible but even undetectable. And only at the second thought it proved that the meter sticks and clocks, standing at rest on duty at the properly chosen observation points in the vicinity of the moving object, do record its true contracted shape as was predicted by Einstein. Very often, though, the word "observer" serves well as a good abbreviation for the set of instruments and simplifies space-time terminology, which is cumbersome as it is. But focusing on the observer as a human being and on his or her opinion, we sometimes are distracted from the instruments by means of which this opinion has been formed, and which, from the standpoint of physics, play a much more important role than the observer himself. Judge it for yourself from the well-known example that follows further.

Let us resort to the twin paradox, which often serves as an identification card of relativity. At the first sight, its only asymmetry is connected with the twin-traveler himself, who reverses the direction of his motion and therefore must be responsible for his paradoxical "rejuvenation" with respect to his brother who stays at home. But on the other

5. CLOSING THE GAPS 267

hand, the twin-traveller, while making his fateful reverse, is supposed to remain safe and sound with his age not undergone any variation. And if even something does happen to his age during the reverse, that something could be made negligible (in comparison with the "rejuvenation") by making the path of his travel as long as necessary. Since the twins themselves do not give us any clue to the asymmetry of their ageing, we have to focus on their measuring instruments.

Each of the twins must have his own set of instruments including his personal clock, which shows his own age, and an array of mutually synchronized spatially separated clocks which are distributed along the route of the travel and show (one clock at a time) the current age of the spatially separated brother. One set of the instruments remains at rest, and the other one moves with the same uniform velocity as the twin-traveler. The personal clock of the traveler behaves similarly to its master and does not bear any substantial asymmetry. If even something does happen to its reading in the process of the reverse, that something could be made negligible (in comparison with its "rejuvenation") by making the path of the travel as long as necessary.

But not so for the moving spatially separated clocks! At the first moment after the reverse they are unable to serve at all, because they have become non-synchronous. To provide the synchronism, all these clocks must be synchronized anew. This operation would be asymmetrical (which vindicates the asymmetry of the traveler's "rejuvenation") and very time-consuming with the time required commensurable with the duration of the travel. This non-synchronism is proportional to the length of the travel and, therefore, cannot be made negligible by increasing the distance covered by the twin. It is an ineradicable source of the asymmetry and it makes the paradox solved, because after the resynchronization, the twin-traveler will change his opinion about his brother's age. In accordance with the new readings of his spatially separated clocks, he will regard himself younger, which will coincide with the opinion of his brother and will be visually justified at their meeting.

To bring the twin story (which is not a paradox now) to its end, the twin-traveler has several options, which will bring him to the similar results. The simplest of them is to make a stop. While being at rest, together with all his clocks, the traveler has all rights to use his brother's spatially separated clocks, which would be indistinguishable from his own clocks after their time-consuming resynchronization. These clocks will tell him about his being younger with respect to his brother. (The "rejuvenation" will be, though, twice less than in the case of returning home with a relativistic speed.) After that he may return home "by train" to avoid any further relativistic complications and to see with his own eyes that his "rejuvenation" is the case indeed. Or he may abandon his spatially separated clocks and return home on the board of his spaceship with the same relativistic speed in a blindfolded manner — being blissfully unaware of the age of his brother until their longed-for meeting.

This instructive story teaches us never to neglect the spatially separated clocks — especially when they are used implicitly. These clocks affect only the opinions of the twins and leave the twins themselves unchanged. Since the reverse (or the stopping) of the twin-traveller makes him to change (or to readjust) these clocks, his opinion also undergoes a relevant change automatically.

As for the twin-traveller himself, it is very hard to think of him as of an elastic creature with all the particles within his body returning to their equilibrium positions after such an intensive perturbation as a relativistic change in the velocity of his motion, when not only the speed but also the Lorentz-factor suffers a considerable change. This means that not only the forces of inertia arise, which are followed by the relevant deformations, but also the Lorentz deformations take place as a reason for the failure of the living tissues. That failure has nothing to do with the forces of inertia and takes place even in the case of a very gradual acceleration when the forces of inertia are negligible. Perhaps Einstein intuitively had felt these complications when suggesting the "clock paradox" as he called it and not the "twin paradox" as was advertised later. Janossy was very near to these complications all the time in his book, which made him very suspicious about relativity, as interpreted by him in a non-traditional way. Actually it was not relativity but only its dynamics that was under suspicion. But, in Janossy's interpretation, the dynamics of relativistic effects was almost identified with relativity itself. Feinberg even blamed him jokingly for "acquitting" relativity in a way practiced in the convictions of the prerevolutionary Russian court: "To acquit but to keep under suspicion" [20]. The traditional relativity, as applied to stationary situations, is, in fact, above any suspicion in contrast to its dynamics, which, strictly speaking, only

5. CLOSING THE GAPS

suggests the result whose accuracy and the range of validity can be established only in terms of non-postulated relativity. The reasons for such functional separation lie somewhere very deep in the bowels of nature and are beyond the limits of special relativity. Perhaps, the relativity based on the postulates demonstrates the properties of the ether, or spacetime, or vacuum, or whatever we call it, while the non-postulated relativity is concerned mainly with "material" bodies which either inhabit that ether or are its derivatives. To make sure of it, it's enough to compare the Lorentz contraction of the rod in two different situations. In the first of them, the rod is set in motion with a relativistic speed in contrast to the second situation in which the rod is slowly turned from a transverse position to a longitudinal one on the platform that never changes its velocity and all the time moves by inertia with a relativistic speed. In the first case, only elastic rods contract in accord with the Lorentz transformations, and even they do it only approximately. As for the second case, **there is no dynamics there** unless the rate of rod's turning becomes commensurable with the speed of light: Any body, or even a group of independent bodies (including the two halves of a broken rod), obey the Lorentz transformations exactly and unreservedly **at any instant of their turning** − otherwise, the principle of relativity would be violated. Such identity in the behavior between everything that happens to be on a speeding platform, **including not only integral entities but even debris**, suggests that all of them reflect the properties of all-pervading spacetime. In other words, it is the ether (alias spacetime) that keeps the standard of World Time while the clocks, being set in motion, reproduce this standard as exactly as they can. This entity is indifferent about uniform velocities of material bodies (unless they exceed the speed of light in vacuum), but fairly discriminating for the accelerations unless they are of a gravitational origin. From the point of view of physics, it is not very important how to call it − ether, spacetime, physical vacuum − anything you like except emptiness so as to affirm its materiality. However, from historical standpoint, it would be perhaps preferable to call it "ether" for two reasons. First, we have to give credit to Einstein for his unnoticed attempt to restore the ether (under its historical name) as a privileged physical system from which all accelerations and especially rotations should be counted [27]. (See the citation given on page 246 of this review.) It is not fair to coin Einstein's image in the eyes of future generations as a disaffirmer of the ether and a destroyer of classical physics – the role he never claimed about or deserved. Newton's law of gravitation was the only essential thing whose range of validity was substantially confined by Einstein's theory of gravitation (which happened beyond the limits of special relativity). Secondly, we have to concede that if not for Einstein's ingenious success in reconciling his two postulates, special relativity all the same would be built (a few decades later) on the base of Lorentz's ether which, in spite of its indifference to uniform velocities, would survive the revolution under his own name assigned to it by Fresnel as early as in the first decades of the 19th century. Isn't it a high time now to restore its reputation at least as an entity from which accelerations should be counted? Why do we discriminate the ether from the inertial mass, which is also undetectable within the limits of the motion with a uniform velocity?

Literature

[1] K.F.Schaffner, The Lorentz electron theory of relativity, *Am. J. Phys.*, **37**, 5, 498-513 (1969)

[2] H.A.Lorentz, *The Theory of Electrons and its Application to the Phenomena of Light and Radiant Heat* (a course of lectures delivered in Columbia University, in March and April 1906), Stechert and Co, New York, 1909.

[3] H.R.Brown, The origins of length contraction, I. The FitzGerald-Lorentz deformation hypothesis, *Am.J.Phys.* **69**, 10, 1044-1054 (2001)

[4] A.Mirabelly, The ether just fades away, *Am. J. Phys.*, **53**, 5, 493 (1985)

[5] A.Einstein, On the electrodynamics of moving bodies, in *Einstein's Miraculous Year: Five Papers that Changed the Face of Physics*, published by D.J.Griffits, 1998

[6] There are so many books and articles with relativity and classical physics opposed to each other that making a satisfactory list of them is a challenge. So, we mention only one of them, whose author, a distinguished scientist and the Post-President of the Institution of Electrical Engineers, did his best to explain the behavior of moving charges and currents in the terms of classical electrodynamics: [E.G.Cullwick, *Electromagnetism and Relativity with Particular Reference to Moving Media and Electromagnetic Induction*, Second edition, Longmans, Glasgow, 1959.] But even he failed to express the unipolar induction in the terms of classical electrodynamics: "The apparent polarization of a mov-

ing magnetized body does not arise in the original non-relativistic electron theory of Lorentz, but is the necessary part of the relativistic specification." This conclusion made him introduce the so-called *"relativistic electrodynamics"*, that in fact is nothing else but electrodynamics of Lorentz where the most difficult derivations are made through Einstein's postulates instead of getting them from the Maxwell equations. As for the curious term "relativistic electrodynamics" (which implies the presence of actually impossible non-relativistic electrodynamics), it is indeed a symbol of the epoch of digesting special relativity, which has lasted for almost a century and is far from being over.

[7] P.W.Bridgman, *A Sophisticated Primer of Relativity, 1983*

[8] H.E.Ives, G.R.Stilwell, An experimental study of the rate of a moving atomic clock, *The Journal of Optical Society of America,* **28**, 7, 215-226 (1938).

[9] H.E.Ives, Historical note on the rate of moving atomic clock, *The Journal of Optical Society of America,* **37**, 10, 810-813 (1947).

[10] H.E.Ives, The measurement of the velocity of light by signals sent in one direction, *The Journal of Optical Society of America,* **38**, 10, 879-884 (1948).

[11] H.E.Ives, Lorentz-type transformations as derived from performable rod and clock operations, *The Journal of Optical Society of America,* **39**, 9, 757-761 (1949).

[12] H.E.Ives, Extrapolation from Michelson-Morley experiment, *The Journal of Optical Society of America,* **40**, 4, 185-191 (1950).

[13] E.G.Cullwick, *Electromagnetism and Relativity with Particular Reference to Moving Media and Electromagnetic Induction,* Second edition, Longman, Glasgow, 1959, This book has already been mentioned in this list with a special remark [6].

[14] L.Janossy, Further arguments on the physical interpretation of the Lorentz transformations, *Uspekhi Physicheskikh Nauk* **62**, 1, 149-181 (1957) (in Russian).

[15] L.Janossy, *Theory of Relativity Based on Physical Reality,* Budapest, 1971 (in English).

[16] G.Builder, Ether and relativity, *Australian J.Phys.,* **11**, 3, 279-297 (1958).

[17] Dewan, M.Beran, Note on Stress effects due to relativistic contraction, *Am.J.Phys.,* **27**, 7, 517-518 (1959).

[18] Dewan, Stress effects due to the Lorentz contraction,

Am.J.Phys., **31**, 5, 383-386 (1963).

[19] E.H.Parker, Elementary explanation of Lorentz-Fitzgerald contraction, Am.J.Phys., **36**, 1, 156-158 (1968).

[20] E.L.Feinberg, Can the relativistic change in the scales of length and time be considered the result of the action of certain forces? *Soviet Physics. Uspekhi,* **18**, 7, 624-635 (1975).

[21] R.W.Brehme, Response to "The conventionality of synchronization", Am.J.Phys., **53**, 1, 56-58 (1985).

[22] N.D.Mermin, Relativity without light, Am.J.Phys., **52**, 2, 119-124 (1984).

[23] M.A.Shupe, The Lorentz-invariant vacuum medium, Am.J.Phys., **53**, 2, 122-127 (1985).

[24] D.Bohm, B.J.Hiley, Active interpretation of the Lorentz "boosts" as a physical explanation of different time rates, Am.J.Phys., **53**, 8, 720-723 (1985).

[25] A.Grunbaum, *"Philosophical Problems of Space and Time"*, New York, 1963.

[26] A. Einstein, Zum Ehrenfestschen paradoxon, Phys. Z., 12, p. 509-510 (1911). *Translated to English as* "On the Ehrenfest paradox. Comment to V. Varicak's paper, in Collected Papers of Albert Einstein, vol. 3, p.378 (Princeton University Press, Princeton, NJ, 1993). *This comment is cited much more rarely than it really deserves (in fact, almost not cited at all).*

[27] A.Einstein, *Sidelights on Relativity*, New-York, 1923, pages 13 and 15,16.

[28] A.Einstein, On the ether, in the book *The Philosophy of Vacuum,* Clarendon Press, Oxford, 1991, p.17-18. (Originally published as "Uber den Ather", *Schweizerische naturforschende Geselschaft, Verhanflungen, 1924).*

[29] V.A.Fock, *The Theory of Space, Time and Gravitation*, Pergamon Press, 1964.

[30] A.Einstein, *Autobiographical Notes,* Centennial edition, in the book *Albert Einstein: Philosopher-Scientist,* Edited by P.A.Schilpp, 1970, pages 52-57.

[31] L.Janossy, Acta Physica Hungary, vol 1, p.391 (1952)

[32] D.Bedford, P.Krumm, On the origin of magnetic dynamics, Am.J.Phys., **54**, 11, 1036-1039 (1986)

[33] J.Terrel, *Phys.Rev.*, **116,** p.1041 (1959).

If you want to learn more about relativity

When preparing this book, the main attention was paid to the basics of special relativity such as the properties of space and time discovered by Einstein. Instead of involving all the numerous and sometimes complicated relativistic effects, no matter how interesting they might be, the author spotlighted only main of them, trying to consider them, though, as deeply as possible. In vain would the reader look for an elaborate list of the relativistic effects, many of which are never even mentioned throughout the book. If a curious reader is not satisfied with the width of the panorama unrolled by the author, he or she can apply to a lot of additional literature which may be found in any scientific library.

There are a lot of excellent and accessible books about relativity written in various styles. They present the basics of the theory on different levels of complexity and are destined for different layers of the readers. These books are so numerous that it is hard to present here the full assortment of them or to try to sort them by categories in accordance with the possible requirements of the readers. But almost all of them need some terminological refinements to be matched up with the non-postulated relativity presented here. These refinements might be regarded as misprints which are listed briefly in the following table:

	Before refining	After refining
1	Special relativity brings about something (for example, the time dilation) which does not follow from pre-Einsteinian physics in any way.	Everything that follows from special relativity can also be derived from classical pre-Einsteinian physics, but no one had known about these corollaries before relativity was developed.
2	Relativistic electrodynamics	Electrodynamics. Non-relativistic electrodynamics has never existed.

	Before refining	After refining
3	Einstein extended Galileo's principle of relativity from mechanics to electrodynamics.	Einstein was first to discover and explain that nothing was to be done in order to extend Galileo's principle of relativity from mechanics to electrodynamics.
4	According to Einstein, the ether wind does not blow not because the bodies suffer the length contraction, but because the properties of space and time are incompatible with that wind.	The length contraction and other relativistic effects are exactly what prevents the ether wind from blowing.
5	Einstein postulated that the velocity of light does not depend on the motion of the observer (or the velocity of light is the same in all the inertial frames of reference).	Einstein postulated that the velocity of light does not depend on the motion of the source. (Neither it depends on the motion of the observer, but this was not postulated – it was deduced from the other Einstein's postulate, according to which all the inertial frames of reference have equal rights.)
6	If a certain body is set in motion with a uniform velocity and then returned back to the state of rest, its length returns exactly to its original value.	Only ideally elastic bodies are able to restore their original size. As for the real bodies, they suffer a residual deformation which may be even quite arbitrary in the case of a plastic body.
7	Einstein disaffirmed the ether as a superfluous and useless notion	Einstein disaffirmed the ether as a superfluous notion in his pioneer work in 1905, but restored it 16 years later.

If, however, the reader is inclined to learn more about the properties of spacetime beyond the limits of special relativity, she or he might be advised to read the perfect books "A Journey into Gravity and Spacetime" by John Archibald Wheeler (1990), and "Black Holes and Time Wraps" by Kip S. Thorne (1995), where the last achievements of physics are presented in a very simple and fascinating way.

Acknowledgements

It is not a common practice to express our gratitude to someone whom we have never met before. However, in spite of that, I can't but mention Yakov Perelman, whose remarkable book "Fun with Mechanics" (providentially bought for me and brought from afar by my dear father as an entertaining substitution for the boring toys) introduced to me the magnificent world of relativity and predetermined my preoccupation with this world which lasted the major part of my life.[1] I am very grateful to my mother Lia (Elizabeth) Plisetskaya-Lomize and my father Grigory Lomize, who were selflessly doing their best to give me a relevant education in the hard times of the war and of postwar period in the former USSR. I am indebted to my unforgettable high school teachers V.B. Shaposhnikov and N.N. Tavdgeridze, who showed me the way to physics. I always remember professors V.A. Fabricant and K.M. Polivanov for their informal presentation of basics of physics and electrical engineering on the professional level. Their lectures were a great support for me — especially their belief in electrodynamics of moving bodies as a universal means for all effects caused by motion. My conversations with them always served me as a source of inspiration for trying to solve the riddles of physics.

This book would have never been written if not for S.L. Nikiphorova, my teacher of literature in high school, who cherished my inclination to writing and was deeply discouraged when she came face-to-face with my unbending striving for physics and technology instead of trying to become a writer as she expected.

When getting out of the traps of relativistic paradoxes, I would have been helpless without long and fruitful discussions with my postgraduate students V.N. Kallagov and V.A. Kuzmin. A lot of thought experiments that we carried out together usually gave us a clue to solve the puzzle.

---

1. As I learned much later, I had been lucky to receive the very first edition of this book, whose last chapter, eliminated by the author from all the numerous subsequent editions, had been devoted to special relativity.

It is hard to overestimate a very productive and friendly exchange of information with professor B.M.Bolotovsky and the constant support generously lent by such outstanding physicists as E.L.Feinberg and V.L.Ginzburg from the Department of Theoretical Physics at Lebedev Institute of Physics in Moscow.

I would have never had the necessary time for getting to the ultimate causes of the relativistic effects but for academician A.L.Mintz – the chief of the Institute of Radio Engineering in Moscow (and the pioneer of high power wireless, radar technology, and charged particle accelerators in the USSR) who agreed to include my group of engineers into his famous scientific and technological team where the passion for science was highly appreciated. I acquired there such highly professional aids and colleagues as Innokenty Sazhin, Boris Rubtzov, Alexander Shmidt, George Doroshenko, and many others who were taking upon themselves a good deal of the current technological routine so as to save my time and energy for the physical research in the field of electrodynamics and special relativity.

Like many other physicists and engineers engaged in current research, I always found it very boring and too time-consuming to pass through the routine connected with the final publication except, of course, regular preprints, which are inseparable parts of the current work. My ideas about the new way of presentation relativity would never have been published if my dear wife Lioudmila Lomize, whose current occupation was connected with arranging various popular scientific lectures all over the country, and my two sons Sergei Lomize and Andrei Lomize, both qualified specialists in physics, had not exerted strong pressure on me so as to urge me to create the first, Russian version of this book and even helped me to grope some ways to publishers in Russia, where all of us lived at that time.[1] The publication became possible thanks to professor G.Y.Miakishev, the official reviewer, who was not only scrupulous in inspecting the whole book but also effective and friendly in making critical remarks which helped me to improve the first version of this book and to make it less controversial. Other pieces of good advice were offered me by professor L.V.Leites, my first cousin, who found time to read the book in spite of his preoccupa

---

1. This first version was written and published in 1991 under the title "From High School Physics to Relativity".

ACKNOWLEDGEMENTS

tion with his research in the field of the physics of high power transformers and reactors.

As for publishing this book in the USA – in the country which became my second homeland in 1997 – it was my cousin Minna Perelman, translator of scientific literature, who not only made the translation of my Russian writings to English, but also helped me a lot with improving my English. Her instructive explanations paved the way for Susan Glovsky, Brenda Millet and Jean Holster – the college teachers in Ann Arbor, Michigan to carry me through a systematic course of English grammar. I am afraid that my progress in English made me too brave and I dared to add accidentally some paragraphs into the book. Therefore, when stumbling on a clumsy sentence, the reader must be sure that it is me who is to blame for it, and not Minna Perelman to whom I am really very thankful.

In the process of preparing this book for the publication, I discovered that I am not lonely in my authorship. My dear son Andrei Lomize – a biophysicist from the University of Michigan – who became an enthusiast for the new way of presenting relativity, made a considerable contribution by simplifying and clarifying my explanations, which made this book much more accessible to the American reader.

I feel obliged to Kristine Willimann – my college teacher, who showed me the way to the basics of computer graphics and self-publishing in the USA. I am also grateful to my grandson Misha Lomize for the pieces of good advise which brought the design of the book much nearer to the American standards, and to Jenny Lomize, my granddaughter, who did her best to help me at the final stage of publishing.

At the final stage of the publication I would be bogged down in the subtleties of the computerized printing if not for my son Sergei, who, living in the other Hemisphere, managed to continuously share his professional experience with me, leaving alone his enthusiastic help in publishing the Russian version of this book.

Index

A

Absolute motion and its imperceptibility 68–69, 84–85, 92, 122, 133–135, 160–169, 171, 198–199
Absoluteness in special relativity 232–233
  event 173–179
    definition of 176
    reality for all the observers 179–193
  interval 176–179
  invariants 176
Acceleration
  as a cause for destroying a solid body 77
  as opposed to a change in velocity 184, 255
Addition of velocities
  See Relativistic law for relating velocities
Ampere's law 21, 25, 41, 57
Associated field
  of a moving charge 56–58
  deformation of 58
  See also Electric field

B

Beam of light 42

C

CGS system of units 18, 21, 24
Change of a variable 46
Charge
  See Electric charge
Classical electrodynamics 44, 273
  connection with mechanics 200
Clock
  tick slowdown
    See Relativistic effects

hand displacement
  See Relativistic effects (time-space dependence)
with a reverberating flash of light 112
with a reverberating flash of sound 114–115
with a spring-driven pendulum 115–116
with a suspended pendulum 116–118
hour-glass 119–120
See also Spatially separated clocks
Clock paradox 120–122, 188–192, 266–268
Compass 30, 55, 174–176, 195–196
Constant of proportionality
  between force and deformation 115
  in Ampere's law 21, 41
  in Coulomb's law 18, 41, 67
  in the expression for a displacement current 39–41
  in the expression for the Lorentz force 24
  in the law of universal gravitation 67, 97
Coulomb's law 18, 21, 41, 50, 51, 57, 67
Crocodile scenario 184–187, 261–264
Current
  See Electric current
Current-carrying loop
  See Magnetic dipole

D

Definition of
  displacement current 38

elasticity of the spring 115
electric charge 18
electric current 21
electric field 19, 31
force 96–97
magnetic field 31
mass 96–97
spatially separated clocks 136
time 111, 269
a physical variable 67, 96, 111
Diffraction 43
Dimensionless velocity 47
Displacement current 38–43, 57, 78
definition of 38
of a moving charged filament 39–41

E

Einstein, A. 11, 65–74
Einstein's
obligation 11, 248–249, 250–251
postulates 11, 13, 65–74, 171, 172–173, 274
as apparently incompatible with each other 171
the second postulate as refined by Einstein 240–245
the second postulate as refined by Bridgman 252–253
Elasticity
of the spring 115
special kind of 182, 183, 245, 251, 255, 263–266, 268–269
Electric charge 67
definition of 18
interaction with another charge 18
density of 44, 61
conservation of 100, 203
moving at a constant velocity 49–58
acted upon by an electric field 19
acted upon by a magnetic field 23
Electric current
definition of 21
density of 21
acted upon by another current 21
created by a moving charged filament 39
convection current 41
inside the atom 30
See also Net current
Electric field 18–20
definition of 19, 31
associated 99
vortical 35, 78, 99, 100
generated by a changing magnetic field 35
of a point charge at rest 19
of a moving charge 73, 78
of two point charges at rest 20
of a charged filament 39
in electromagnetic wave 41–43
varying 38, 78
deforming a solid body 76
exerted upon a charge 18, 79, 81
energy of 20, 101
lines of force of 19, 33, 38, 100
Electrical engineering
voltage 28, 30
resistance 30
electromagnetic induction 28, 34, 36
electric motors as based on the Lorentz force 27, 28
rotating generator as based on the Lorentz force 28, 37

INDEX

281

rotor 27
stator 27, 28
brushes 27, 28
collector 27
electrical transformers 37
unipolar generator 204–206
Electromagnetic
   field 59, 62
      lines of force of 41–43
      contribution to the sound wave 115
      of a moving charge
         See Electric field
   induction 28, 36, 78, 99
   law of 34, 73
   mass 97–100
   nature of light 41
   radiation 58
   wave 41–43, 90
Electromotive force 37
   external 36
   of self-induction 36
   caused by electrification of a rotating magnet 204–206
Energy
   kinetic 95, 104
   of electric field 20
   rest energy 104
Equilibrium
   between the parts of a body 62–66, 76, 78
   position 76
   stability of 76
Equivalence between inertial and gravitational mass 96–97
Ether 234
   invisibility in the case of uniform velocity 68–73, 91, 109, 114–115, 135, 156, 160–169, 198, 203–204, 232
   as a hypothetical medium that will fade away later on 11, 32, 33, 82, 83, 89, 90, 199, 237–238
   in Einstein's early interpretation 246
   in Einstein's mature interpretation 246–247
   as a historical reminder 31–32, 38, 155–156, 170–171, 239
Experiment of Ives and Stilwell 72, 253

F

Faraday, M. 34
Feinberg, E.L. 14, 257, 268, 272
Field
   See Electric field, Magnetic field, Associated field
Force
   definition of 96–97
   as a cause for acceleration 95
   of elasticity 116
   of gravitation as caused by an acceleration in an inertial frame of reference 212–215
   of interaction between currents 21, 76
   of magnetic attraction 25, 81
   produced by electric field 18, 19, 25
   produced by magnetic field 24
   projections of 63
   rotating the rotor of an electric motor 27
   transformation 62
   See also Coulomb's law and Ampere's law
Frame of reference 47
   inertial 209–212
   laboratory 68
   proper 68

G

Galileo 69
Galileo's principle of relativity 69, 71, 77, 86, 170, 237,

239, 240, 242, 274
Gravitation 96
  a riddle to be solved 220–222
  absoluteness in the universe 233–235, 247
  as a cause for bending light rays 215–216
  curved space 224–229
  curved spacetime 230–231
  Einstein's theory of 233
  free fall as a kind of free hovering 211–212, 217–218
  propagation of 224
  red shift 216–217
  revision of the Newtonian force of gravity 218–220
  speed of propagation 229–230
  strange tickle 222–224
  vs. general relativity 233

I

Induction
  See electromagnetic induction
Inertia 36
Ives, H.E. 253

J

Janossy, L. 14, 254, 255, 259, 268, 271

L

Law of electromagnetic induction
  See Electromagnetic induction
Lenz's rule 36, 99
Light
  See Beam of light, Speed of light
Light barrier
  See Speed of light
Light-ray clock vs. sound-wave clock 114
Local time
  as a fictitious variable 60
  as a true variable 132–135
Lorentz, H.A. 100, 59–73
Lorentz contraction
  of a solid body 62–64, 76–94, 100
  of an associated field 54–56, 59, 73
Lorentz factor 47, 56, 104
Lorentz force 25, 26, 28–31, 37
  physical meaning of 31–33
  saving the electric motor from rotating too fast 28
  turning the pointer of a compass 31
Lorentz transformations 44–48, 49, 50, 70, 74
  mathematical representation of 47
  as a means for solving the Maxwell equations 45–47, 59–61
  as leading to
    length contraction 82–84
    clock tick slowdown 122
    clock hand displacement 149–150
    time zones on a moving platform 123–124
    independence of the speed of light from the motion of an observer 158
    electrification of a moving current-carrying loop 200–202
  in Lorentz's interpretation 67, 256
  in Einstein's interpretation 67, 69, 172, 256
  in Janossy's interpretation 255–256
  symmetry of 48, 72

M

Mach-Einstein principle 234–235
Magnetic dipole 60
Magnetic field 21–33

INDEX

definition of 31
as a cause for inertia 36
as a vector 22
as compared with electric field 23
as created by a current 22
as creating a voltage in a rotating generator 28
between the plates of a moving capacitor 195
exerted upon a current-carrying frame 27
exerted upon an electric charge 23, 79, 81
flux of 36, 98
in electromagnetic wave 41–43
lines of 22, 25, 33
non-uniform 26
of the Earth 22
propagation vs motion of 33
varying 78
Mass
definition of 96–97
gravitational 67, 96–97
inertial 95–110
longitudinal 108
of a resting body 104
origin 97–100
transverse 108
Mass-energy equivalence 71, 102–105
Maxwell, J.C. 38, 44, 72
Maxwell equations 44–48, 59
remarkable property of 46
Measurement
as the only way to a physical definition 111
of the instruments at rest by means of the instruments in motion 160–169
as applied to a rod 167–168
as applied to a single clock 163–166

as applied to spatially separated clocks 162–163
of the length of a moving rod 68
Measuring instruments 66–68, 70, 71
as a reference from which motion has to be counted 31
laboratory 129
proper 67
symmetry between motion and rest for 160–169
See also Observer
Michelson's experiment 62, 72, 73, 86–92, 113
acoustic analogy of 88
as a conspiracy of the instruments to hide the ether 90
radar analogy 86
Michelson's interferometer 90–91

N

Net current 40, 41
Newton, I. 72, 105
Newton's laws 71, 72
the first law 57
the second law 105, 106
Non-primed variables 47, 49, 59, 62, 67

O

Observer 66, 67, 68, 82
moving against a flying body 156–158
moving against a ray of light 151–155
who overslept the acceleration 182, 183
with all his instruments working from a single point 266
See also Terrel effect
See also Measuring instruments
One-way speed of light

experiments 154

P

Permanent magnet 23, 25
Piezoelectric 77
Poincare, J.H. 61
Postulates
 See Einstein's postulates
Primed variables 48, 49, 62, 67, 72, 74
Principle of equivalence 77, 233
Principle of relativity
 See Einstein's postulates

R

Relativistic dynamics 105–110
 the first rule of 106–107, 116
 the second rule of 107–108
 the third rule of 108–110
  as a good reason for rotating a moving current-carrying loop 207–208
Relativistic effects 11, 65, 74, 274
 length contraction 59, 62–64, 70, 73, 76–94, 100
 in a human eye 84, 85
 mass-velocity dependence 71, 100–102, 105
 clock tick slowdown 70, 72, 111–122
  experimental proof 121–122
  general rule for 120, 122–123
  in an hour-glass 119–120
  in the clock with a spring-driven pendulum 115–116
  in the clock with a suspended pendulum 116–118
  in the light-ray clock 113
  in the sound-wave clock 114–115
 reality of 179–181
 time-space dependence 123–150
  as a result of a slow transportation of a clock on a moving platform 70, 129–137, 252–253, 258–259
  as caused by a change of the frame of reference 133–135, 136–137
  as caused by acceleration of connected clocks 139–147, 148, 183–187, 259–264
 force transformation 62, 79–82
 electrification of a moving magnetic dipole
  explained in terms of classical electrodynamics 194–199
  explained in terms of classical mechanics 199–200
 electrification of a current-carrying wire
  at rest 202–203
  in motion 202
 electrification of a rotating current-carrying rim 203
 electrification of a moving magnet 203–206
  taking place at ordinary velocities 194–206, 208
  dynamics of 85–86, 139–147, 181–182
"Relativistic" electrodynamics 271, 273
Relativistic law for relating velocities 157–158
Relativity of simultaneity 129, 136–137, 172, 175, 238, 242, 252, 257, 258–269
 parable about absolute time 137–139

S

Self-induction 36
Slow transportation of a clock on a stationary platform 125–

INDEX

129
   on a moving platform 129–137
Solid body 76
Sound propagation 115
Spacetime
   See Ether
Spatially separated clocks 65
   definition of 136
   synchronized by slow transportation 65, 125–135, 258–259
   synchronized with light signals 65, 127, 136
   synchronized with mechanical signals 147
   without synchronization 147–149, 259
   rod-clock analogy 148
Special relativity 72–74
   and classical physics 11, 69, 71, 139, 170, 237–239, 249–250, 270–271, 273
      erecting the bridge between 239–240, 248–249, 252–269
   inapplicability to non-equilibrium systems 148, 182–183
   postulates
      See Einstein's postulates
   vs general relativity 233
   and nuclear energy 104
   as a cozy island surrounded with Unknown 234–235
   in four-dimensional form 71
Special relativity in a non-postulated form
   first steps 237
   Einstein's contribution to 244–245, 248–249
   Bridgman's contribution to 252–253
   Ives's contribution to 253–254
   Janossy's contribution to 254–257
   Feinberg's contribution to 257
   deriving first-order effects from classical physics 257–269
   as a means for solving relativistic paradoxes 192–193
Speed of light 21, 41, 66, 70, 83, 90, 115
   with respect to the moving observer
      as measured by the instruments at rest 151–153
      as measured by the instruments in motion 153–155
   a fantastic excursion with a superlight speed 93–94
   as an unreachable barrier 93–94, 101–102, 108
   as speed of gravitational disturbances 229–230
   disappearing from equations 103
   independence from the motion of the observer 155–156
   independence from the motion of the source 69
   principle of constancy of 155, 240–242
   influence on the speed of sound 115
Speed of sound 115
Stilwell, G.L. 253

T

Taylor, E.F. 225
Terrel, J. 85, 266, 272
Terrel effect 85, 266
Thermal oscillations 76, 77, 82
Thorne, K.S. 174, 274
Time

definition of 111, 269
hidden in the bowels of the ether 111, 269
Newtonian vs Einsteinian 249–250
parable about absolute time 137–139
zones on the moving platform 68, 132–135, 182, 184–187
See also Relativistic effects
Twin paradox
See Clock paradox

U

Unipolar generator 204–206
Units of measurement 67
See also CGS system of units

W

Wave 33, 41–43
of sound 114
See also Electromagnetic wave
Wavelength 42, 90
Wheeler, J.A. 225, 247, 274
Work 95